A Concise Introduction to Scientific Visualization

Brad Eric Hollister · Alex Pang

A Concise Introduction to Scientific Visualization

Past, Present, and Future

Springer

Brad Eric Hollister
Department of Computer Science
California State University,
Dominguez Hills
Carson, CA, USA

Alex Pang
Department of Computer Science
University of California, Santa Cruz
Santa Cruz, CA, USA

ISBN 978-3-030-86418-7   ISBN 978-3-030-86419-4   (eBook)
https://doi.org/10.1007/978-3-030-86419-4

Main Image/Drawing:
• The water mechanism/Leonardo Da Vinci: © wowinside/stock.adobe.com

Image Bar from left to right:
• Simian immunodeficiency virus, 3D model. © Donald Bliss (Nlm), Sriram Subramaniam/National
Cancer Institute/Science Photo Library
• RBC, SEM: © DENNIS KUNKEL MICROSCOPY/Science Photo Library
• Rayleigh Bernard Convection simulation: (CC BY-SA 3.0): https://commons.wikimedia.org/wiki/File:
RayleighBernardConvection.png/http://wiki.palabos.org/community:gallery:rb_3d/Author: LBMethod.
org/jonas

This Springer imprint is published by the registered company Springer Nature Switzerland AG
The registered company address is: Gewerbestrasse 11, 6330 Cham, Switzerland

Preface

Much of our cortex is dedicated to processing visual information, and it imposes upon mental models a physical intuition. While there are exceptions in modern science when our visual nature leads to artifact, for the vast number of problems our ability to visualize the natural world has elevated human understanding to its current level. Using imagery, we are able to see the unseeable, and thus further knowledge.

This treatise will outline scientific visualization. While there is an established related field of information visualization, we mostly address visualization for scientific purposes. Despite modern buzzwords like "data science" and computer graphics in popular media, the general public (and even some academics) remain unfamiliar with scientific visualization. But, anyone capable of mental pictures, often instinctively uses visualization to solve problems they encounter. If their solutions model the natural world, or produce useful abstractions from it, then this type of problem-solving is considered scientific visualization.

Scientific visualization is not an isolated area of research. While today, visualization is primarily computer-generated, visualization in science stretches back to a time well before computers! That said, we do not consider problems in the broader discipline of visualization that are not scientific. Nor do we consider areas of study such as realism in art. When illustration relates to scientific visualization, it is discussed to convey context.

The first two chapters cover the role of geometry in natural science and scientific visualization. A link is drawn between Euclid and the work of da Vinci (and others) of the Renaissance period. Then, the kinematics of celestial motion is presented in connection with later methods for shape and curvature description. Starting with chapter three, we describe Faraday's insight into invisible electromagnetic fields. Faraday was known to have had his great revelation through visualization of the phenomena. As another case study, Lawrence Bragg, a scientist known to possess an early aptitude for spatial problems, contributed to molecular visualization and the first direct experimental structural determination of matter. Computers still had not been invented yet during this era of scientific visualization, but that was soon to change. In the last two chapters, modern scientific visualization starts to take form. We see how early computer use was directed exclusively at problems of science.

However, not until the latter part of the twentieth century, did the computer become sophisticated enough to draw interactive imagery.

Carson, USA   Brad Eric Hollister
Santa Cruz, USA   Alex Pang

Contents

Early Visual Models

Abstract This chapter serves as an entry point into our introduction of *scientific visualization*. The first externalizations of visual thought are considered [1]. The focus is on the work in the pre-Enlightenment before the middle ages, which are discussed in Chap. 2 as a precursor to the Renaissance. This period's drawings were crude comparatively, but suggestive of the future trajectory of human expression and cognition [2]. The time span is also vast, greater than what is covered from the other chapters combined. However, prehistory is sparse with example. It is most significant when viewed as the watershed before systematic analytical thought [3, 4]. The transition from prehistory to Greek mathematical philosophy [5], especially regarding Euclidean geometry, was key to the later developments in perspective theory and the scientific method.

Prehistory

From the vantage point of modern culture and technology, it is difficult for us to directly relate to the perspective and actions afforded those of considerably early times. It was generally the case that most visual artifacts were produced as records of the world, and less of mental imagery and ruminations on reality.

Unless, however, we consider prehistoric "art." Then, by discrepancies between prehistoric drawings juxtaposed with rational modern representation (perspective, proportion, etc.), we begin to see what is considered deficiencies in our ancestors' own internal view and today's conceptualization of the external world. Analyzing these discrepancies lead us to an estimation of what was visualized (the products of earlier minds) in the prehistoric period.

Another running theme of the prehistoric period is the lack of delineation and intent between the various forms of visual invention. As there was no "science," one can not interpret this period's achievements as directly in service of *scientific* visualization.

We will see later how these symbolic inventions became general forms of expression which were to be later built upon and shaped into more specific tools for scientific endeavor.

B. E. Hollister and A. Pang, *A Concise Introduction to Scientific Visualization*,
https://doi.org/10.1007/978-3-030-86419-4_1

Because of visual tools, namely glyph (i.e., carving) and art, subsequent generations were able to extend and elaborate upon prehistoric visualization, using ideas enabled by the earlier foundations.

Parietal Art

In southern France, approximately 30000 BCE, was the site of the earliest known petroglyphs (stone carvings) and parietal art, see Fig. 1. Many of these works also contained additional repetitive markings superimposed over animal imagery.

It can only be speculative, but the prevailing notion is that these glyphs were meant to contain information (number of animals in a herd, etc.), akin to information linked to spatial representation [6]. No speech was recorded yet, as this was well before phonetic writing.

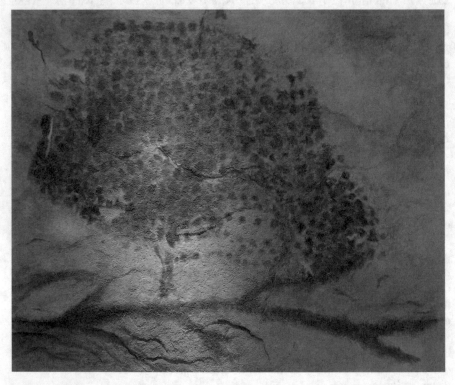

Fig. 1 Cave painting and engraving from Marsoulas in France approximately 30,000–32,000 years ago. From [HTO, Wikimedia Commons]

Fig. 2 Non-political view of the Tigres and Euphretes rivers. From [DEMIS Mapserver, Wikimedia Commons]

The Near East

The invention of symbolic representation itself was a key accomplishment. Symbolism represented itself most prominently with the emergence of script. The most successful prehistoric societies were successful in their ability to extract from reality defining features and externalize them; for communication, expression, and by consequence planning and understanding.

The earliest records of this come from Mesopotamia, between the Tigres and Euphretes rivers (the "Fertile Crescent"), often considered the "cradle of civilization." See Fig. 2.

Proto-Writing

The first known forms of proto-writing (or proto-cuneiform) were clay "tokens" dating back to around 8000 BCE. These tokens consisted of small geometric objects such as cones and spheres (see Fig. 3) [7]. They are considered by archeologists to have been used for accounting in trading and commerce. It is likely that the geometric forms represented items of inventory, symbolized by basic shape.

Fig. 3 Clay bulla, or
"bubble," with token
contents displayed. From
[Marie-Lan Nguyen,
Wikimedia Commons]

Pictographs

As we will see, the concept of glyph is today used throughout scientific visualization
as a nonverbal symbolic representation of a quality or quantity (arrow and bar glyphs,
for example) [8] as shown in Fig. 4.

Thus, the modern practice of scientific visualization can be considered as pictorial
symbols, or pictographs (also know as "pictograms"). In this text, we will not explore
how pictograms evolved into phonetic representation, as we are interested in the
historical development of the practice of visualization for scientific inquiry.

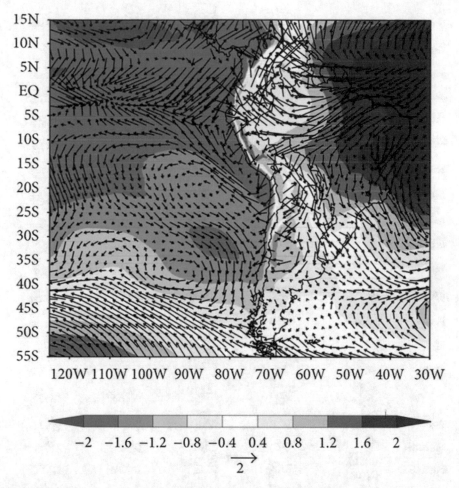

Fig. 4 El ñino effects in South American climate as represented with wind force directions in a vector field plot using arrow glyphs in modern scientific visualization. Reproduced from [South American Climatology and Impacts of El Niño in NCEP's CFSR Data, Hindawi Limited, Open Access 2013]

The first use of pictograms given current evidence, [9] was in the cities of Sumar and Uruk circa 3000 BCE. The pictograms (Fig. 5) consisted of concrete and abstract concepts, in the form of line drawings impressed upon clay tablets using reeds (cuneiform engraving).

A key step in the human visual system is to detect edges through value change in objects [10]. Pictograms resemble this decomposition of form with proportional invariance (or singular perspective).

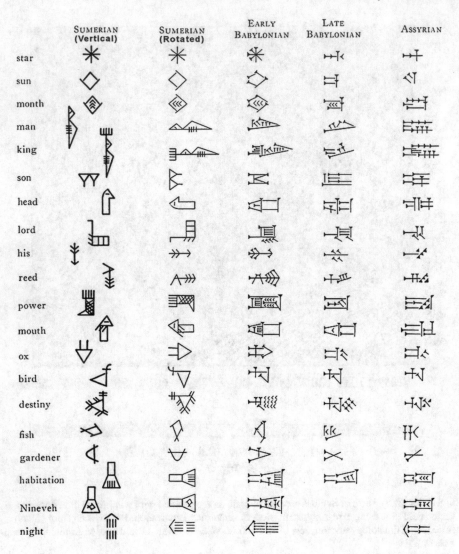

Fig. 5 Original Sumarian to Assyrian cuneiform pictograph. Leftmost column provides translation. Reproduced with permission from [A History of Writing by Albert Mason, Wikimedia Commons, 1920 Public Domain]

Counting and Numerals

Perhaps the most important of all abstractions is the notion of count [11]. Implied in this, is the idea of a set of objects sharing features or possessing commonality. As pictograms are abstractions of form and object, numbers are abstractions for quantity and object similarity. Naturally, as with symbols for various types of animals or land features, notations were extended to represent numbers.

From proto-writing (the clay bulla in Fig. 3) followed clay tablets and systems of record keeping derived from proto-cunieform tokens [12, 13]. An example circa 500–3350 BCE from Uruk is shown in Fig. 6.

Greece

Ancient Greek culture contributed to the foundations mathematics. This is no more evident than in *Euclidean* Geometry. There is some criticism as to the degree of its success, and how it fell short of projective geometry (used in realistic visual

Fig. 6 Tablet digits are listed left-to-right in decreasing order of significance. From [Poulpy, Wikimedia Commons]

Fig. 7 The "Cubiculum" fresco painting, from 50 to 40 BCE. Note the lack of proper perspective, as there is only the approximation of a vanishing point and irregular foreshortening. Not until the Renaissance, was one-point perspective formalized. From [Cubiculum (bedroom) from the Villa of P. Fannius Synistor at Boscoreale, The Met]

depictions [14]). Nonetheless, the method of logical argument (proof) in geometric constructions laid groundwork for later developments.

This distinction is important, however, as the Greeks never produced paintings which fully accounted for perspective. A sample from the Hellenistic period, [15] the period of Greek culture after Classical Greece, is shown in Fig. 7. Here, we see that the building structures are without a common vanishing point.

The Beginnings of Geometry

Euclid of Alexandria lived from 323 to 283 BCE. He produced *Elements*, laying the foundation of geometry, mostly in "flat" two-dimensional space. Part of the original work is shown in Fig. 8.

The material in *Elements* is familiar to most from secondary school. *Elements* represents the initial systematic postulating of theorems, deduced stepwise from axioms (fundamental observations). Using visualization and measurement as justifications in derivation, *Elements* describes a consistent visual model (set of abstractions) for *flat* shapes such as circles and polygons—themselves being axiomatic generalizations of physical objects (Fig. 9).

In the last book of *Elements* (Book Eight), Euclid describes mathematically the Platonic Solids, three-dimensional geometries composed of plane polygons. This treatment was expanded upon by Leohnard Euler, via *Euler's Formula* in the sev-

Fig. 8 A remnant of Euclid's *Elements*, found at Oxyrhynchus. The figure appears in Proposition Five of Book Two. Reproduced with permission from [Jitse Niesen, http://www.math.ubc.ca/~cass/ Euclid/papyrus/tha.jpg, 2006 Public Domain]

enteenth century ADE, thus beginning the field of topology and graph theory—two significant contributions to scientific visualization discussed in Chap. 2. Graph theory has had a overarching impact on computer science, by treating only quantity and connections of vertices with edges without regard to their spatial locality or proportions. This description is important for areas such as data structures, communication, and neural networks.

As we will see in Chap. 4, modern computer rendering systems use geometries composed of many polygons to approximate arbitrary curved surfaces, for use in graphical scientific visualization applications [16]. This method is also used in finite-element analysis for simulation of continuum material (see Fig. 10).

Visual Methodology

It should also be noted that Greeks used *visual* methods rather than algebraic (or algorithmic) techniques in proof and area calculation. The Pythagorean theorem is shown in Fig. 11. It wasn't until Descartes (Cartesian coordinates, Chap. 2) that geometry could be generalized as an algebraic interpretation. A summary of the visual methods of the Greeks is stated as [17]:

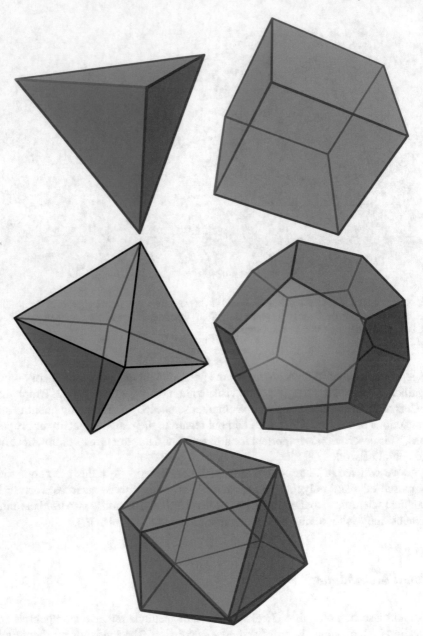

Fig. 9 The Platonic Solids: Tetrahedron, Cube, Octahedron, Dodecahedron, and Icosahedron. From [Kjell André, Wikimedia Commons]

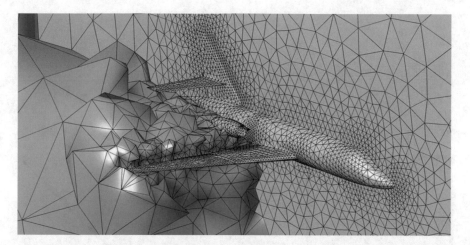

Fig. 10 Example of modern use of polygonal surfaces. Compressible steady turbulence over wing of aircraft. Surface discretization view. Reproduced with permission from [Per-Olof Persson, DistMesh: High-Quality Unstructured Mesh Generation for Implicit Geometries—https://crd.lbl.gov/departments/applied-mathematics/math/software/, 2015]

> With regard to moderns speaking of golden age geometers, the term method means specifically the visual, reconstructive way in which the geometer unknowingly produces the same result as an algebraic method used today. As a simple example, algebra finds the area of a square by squaring its side. The geometric method of accomplishing the same result is to construct a visual square. Geometric methods in the golden age could produce most of the results of elementary algebra.

Indeed, even the Greek word "hypotenuse" means literally, "stretched against." A separate instance, triangle congruence was determined by Greek visual method as well. Ivins explains:

> The way that Euclid proved his basic theorem–that two triangles, having two sides and the angle between them equal, are equal to each other–was by picking one triangle up and superimposing it on the other. [14]

Lo Shu Squares and Visual Methods

The Chinese Lo Shu Square appeared circa 570 BCE, but has been attributed to Fuh-Hi (2858–2738 BCE). An example is shown in Fig. 12.

Some authors such as Peddie, [18] suggest that Lo Shu (or "magic") squares had a significant role in methods of visualization, drawing attention to the similarity of the square and matrices used in linear algebra. Matrices (further generalized as tensors) are arrays of numbers used as conveniences for computing affine transformations in multiple dimensions—among other uses. Interestingly, a modern form of the Lo Shu Square called *Geometric Magic Squares*, [19, 20] created by Lee Sallows in 2001,

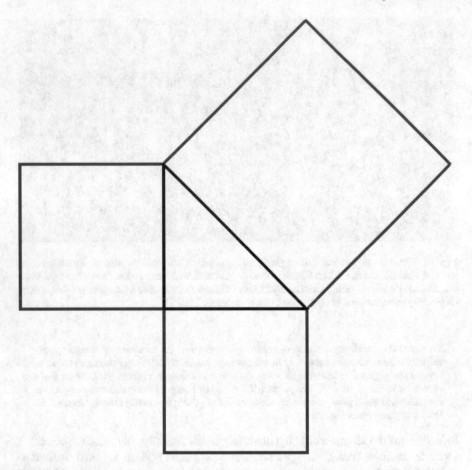

Fig. 11 Here, we see a visual method for proving or deriving Pythagorean theorem, which was used to show the areas produced by the red squares could be made to be congruent with the blue square, where each square is a visual representation of the "squaring" of each of the two legs of the triangle. From [Winterheart, Wikimedia Commons]

replace numbers with shapes and geometric operations. As we shall see, algebra appeared much later than the original Lo Shu Square. Algebra's use in geometry followed only later still (Fig. 13).

Conics

The first known mathematical use of conic sections are credited to Menaechmus, who lived from 380 to 320 BCE. Conics were employed in the visual (geometric) solution

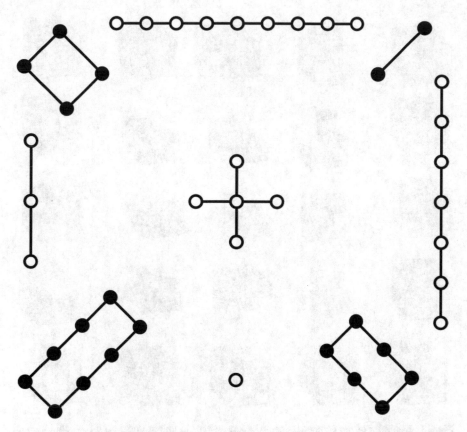

Fig. 12 The Lo Shu square. Each row, column, and diagonal sums to a total of fifteen. From [AnonMoos, Wikimedia Commons]

of the "doubling the cube" problem [4]. One formulation required the simultaneous solution for x and y, with known constants a and b in the following:

  i. $x^2 = ay$;
 ii. $y^2 = bx$;
iii. $xy = ab$.

Menaechmus referred to the parabola (Eqs. i and ii) as a "section of a right-angled cone," and the hyperbola (Eq. iii) as a "section of an obtuse-angled cone." It was found later, by Apollonius of Perga (240–190 BCE), that both could be generated by the intersection of a cutting plane and a right-angled double cone (including circular and elliptical curve intersections) as seen in Fig. 14.

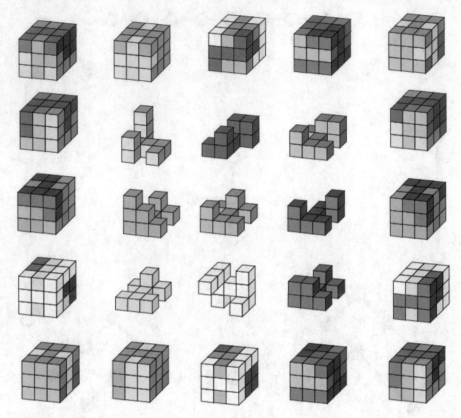

Fig. 13 Exemplar of the geometric magic square, where in this example each row, column, and diagonal combine to form a 3 × 3 × 3 cube of smaller unit cubes. Colors represent individual substructures in dissembled and assembled configurations. From [Lee Sallows, Wikimedia Commons]

Optics

Beyond *Elements*, another work by Euclid that is prominent in early scientific visualization is *Optics*. Euclid's *Optics* was the first major work on the geometry of vision. The model of optics that Euclid proposed [21] was based on the following postulates or axioms without consideration of their physical nature. To illustrate the importance of these axioms, we outline and illustrate them as follows:

1. That rectilinear rays proceeding from the eye diverge indefinitely.
2. That the figure contained by a set of visual rays is a cone of which the vertex is at the eye and the base at the surface of the objects seen.
3. That those things are seen upon which visual rays fall and those things are not seen upon which visual rays do not fall.
4. That things seen under a larger angle appear larger, those under a smaller angle appear smaller, and those under equal angles appear equal.

Fig. 14 The parabola and hyperbola conic sections. From [JensVyff, Wikimedia Commons]

5. That things seen by higher visual rays appear higher, and things seen by lower visual rays appear lower.
6. That, similarly, things seen by rays further to the right appear further to the right, and things seen by rays further to the left appear further to the left.
7. That things seen under more angles are seen more clearly.

Euclid considered the eye to be the source of visual illumination, not light as we know it today. Note that (backwards) recursive ray tracing, the foundation for modern day *Monte Carlo Path Tracing*, a computer-based rendering algorithm that traces light rays backwards from the eye to light sources [22].

An example from Euclid's *Optics* is the Remoteness Theorem, which explains how points farther from an observer appear closer in the view plane. See Fig. 15.

Greek Astronomy and The Antikythera Mechanism

Astronomy is important as orbital paths can be compared to the static geometries of circular and elliptical curves, both familiar conic sections. In Fig. 16, we see the computation of the sizes of celestial bodies by Aristarchus of Samos (310–230 BCE). This visualization lead to *kinematic analysis* of machines as we will see in Chap. 2.

An ancient Greek mechanism was discovered in 1901, near the island of Antikythera. According to Derek J. de Solla Price in 1951 [23], "The purpose was to position astronomical bodies with respect to the celestial sphere, with reference to the observer's position on the surface of the earth." It is perhaps the earliest know predecessor to modern computing machines.

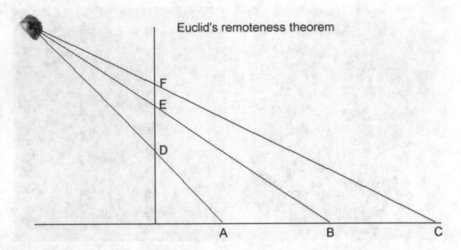

Fig. 15 Euclid's Remoteness Theorem from *Optics*. Note that while points F and E are close on the visual plane, they correspond to points B and C which are farther apart than points A and B (from:). Reproduced with permission from [Otto B. Wiersma, https://www.ottobwiersma.nl/philosophy/perspect.htm, 2008 Public Domain]

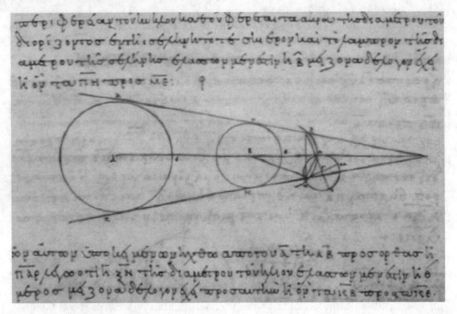

Fig. 16 Sizes of the Sun, Earth, and Moon, from a tenth-century copy. From [Library of Congress Vatican Exhibit, 2006]

Fig. 17 Reproduction of the Antikythera Mechanism. From [Giovanni Dall'Orto, Wikimedia Commons, 2009]

The engineering skill that contributed to its creation is remarkable. It featured bronze gears shaped as equilateral triangles, unfortunately less efficient at transferring smooth motion as the involute gear geometry devised by Christiaan Huygens in the seventeenth century (discussed in Chap. 2) (Fig. 17).

Summary

This chapter covered the vast period from prehistory to the end of ancient Greece. Science had not yet formed as an empirical discipline. But, much visual groundwork was laid during this time for later developments in scientific visualization. Both Euclid's *Elements* and *Optics* were arguably the most suggestive works of this era, especially when considering how the Greeks used visual methods to relate geometry with number—namely the method of *Application of Area* (recall Fig. 11) found in *Elements* [24]. The generalization of computation by algebra and combined with geometry, known as the analytical geometry created during the Enlightenment, will be introduced in the next chapter.

References

1. Robin, H.: The Scientific Image: From Cave to Computer. Abrams (1992)
2. Rovida, E.: Machines and Signs: A History of the Drawing of Machines, vol. 17. Springer Science & Business Media (2012)
3. Klein, J.: Greek Mathematical Thought and the Origin of Algebra. Courier Corporation (1992)
4. Heath, T.L.: A History of Greek Mathematics, vol. 1. Clarendon (1921)
5. Heath, T.L. et al.: The Thirteen Books of Euclid's Elements. Courier Corporation (1956)
6. George, A.: Code hidden in stone age art may be the root of human writing. New Sci. (2016)
7. Schmandt-Besserat, D.: An archaic recording system and the origin of writing. Syro-Mesopotamian Stud. (1977)
8. Eichler, T.P., Londoño, A.C.: South American climatology and impacts of El Niño in NCEP's CFSR data. Adv. Meteorol. (2013)
9. Robinson, A.: The Story of Writing. Thames & Hudson (2007)
10. Hubel, D.H.: Eye, Brain, and Vision. Scientific American Library/Scientific American Books (1995)
11. Mazur, J.: Enlightening Symbols: A Short History of Mathematical Notation and its Hidden Powers. Princeton University Press (2014)
12. Nissen, H.J., Damerow, P., Englund, R.K., Englund, R.K.: Archaic Bookkeeping: Early Writing and Techniques of Economic Administration in the Ancient Near East. University of Chicago Press (1993)
13. Ebrahim, A.: The Mathematics of Uruk and Susa (c. 3500–3000 bce). http://www.mathscitech.org/papers/ebrahim-2019-Mathematics_of_Uruk_and_Susa_3500-3000_BCE.pdf
14. Ivins, W.M.: Art & Geometry: A Study in Space Intuitions. Courier Corporation (1964)
15. Stewart, A.: Art in the Hellenistic World: An Introduction. Cambridge University Press (2014)
16. Morganyand, K., Perairez, J.: Unstructured Grid Finite Element Methods for Fluid Mechanics (1997)
17. Apollonius of Perga. https://en.wikipedia.org/wiki/Apollonius_of_Perga
18. Peddie, J.: The History of Visual Magic in Computers. Springer, Berlin (2013)
19. Sallows, L.: Geometric magic squares. Math. Intell. **33**(4), 25–31 (2011)
20. Sallows, LCF.: Geometric Magic Squares: A Challenging New Twist Using Colored Shapes Instead of Numbers. Courier Corporation (2013)
21. Lindberg, D.C., Lindberg, D.C.: Theories of Vision from Al-Kindi to Kepler. University of Chicago Press (1981)

22. Jensen, H.W.: Realistic Image Synthesis Using Photon Mapping, vol. 364. Ak Peters Natick (2001)
23. Freeth, T., Jones, A., Steele, J.M., Bitsakis, Y.: Calendars with olympiad display and eclipse prediction on the antikythera mechanism. Nature **454**(7204), 614–617 (2008)
24. Boyer, C.B.: History of Analytic Geometry. Courier Corporation (2012)

Illustration and Analysis

Abstract Following the middle ages, the fifteenth through seventeenth centuries marked the Renaissance period where art and literature flourished. Emphasis shifted to reasoning and scientific developments in the Age of Enlightenment, lasting until the start of the nineteenth century. Interestingly, the term "enlightenment" itself suggests the casting of light, or the enabling of visibility, where there was none prior.

This chapter first covers the evolution of scientific illustration from the Middle Ages to the Age of Enlightenment. After this, the history of analytic and descriptive geometry is discussed, both having central significance to the eventual maturation of scientific visualization later in the twentieth century. Finally, we continue our theme for each period described so far, by concluding with the essential step forward in computation. In so doing, we introduce a device called the Pascaline, which has lead to the computers of today—essential for the modern field of visualization.

Middle Ages

Formal mathematical research had receded after the Hellenistic period, but consequently, there was increased interest in observational and graphical methods. The beginning of the seventeenth century would bring with it new mathematics enabling modern science and eventually computational geometry [1].

Modern scientific visualization can trace its roots to advances in mathematics and science. Medieval mathematics was influenced not only by the graphical elements (i.e., the constructions) of Greek geometry but also the ambition to combine geometry with systematic algebra—in the study of solutions (points) associated with equations containing more than one variable [2]. This was to culminate in the seventeenth century with the introduction of analytic geometry.

The oldest known record of using a graphing technique for scientific data was published in the tenth century. Planetary orbits seen in the night sky were tracked and plotted using rectilinear coordinates [3], as included in an appendix (entitled, *De*

Cursu per Zodiacum) from *Macrobius Boetius in Isagog*—a treatise for monastery education. See Fig. 1.

A thirteenth-century design and illustration of a "perpetual motion" machine is shown in Fig. 2. This rendering is reminiscent of an orthogonal view using only lines without lighting characteristics.

Printing and Treatises

As mentioned, a new type of literature was created during the latter part of the medieval period called the treatise. A distinctive quality of treatises were their inclusion of rich illustration to convey mechanical devices, naturalistic art, and more generally, scientific information. These works were enabled by the printing press, invented by Johannes Gutenberg circa 1440 CE. A recreation of the original printing press is shown in Fig. 3.

Scientific Illustration

Some of the first technical drawings printed in treatises used a simplified style. Relevant elements of Rovida's characterizations [4] of technical and scientific illustration from the Renaissance are:

1. Exploded views to illustrate the interplay of multi-part mechanisms. See Fig. 4.
2. Measurements within diagrams, as shown in Figs. 5 and 6.
3. Cutaways and internal views via transparent layers to reveal internal arrangement. Examples are shown in Figs. 7 and 8.

Additionally, a description of the camera obscura by Gaspar Schott (1608–1666 CE) was first described using illustration. See Fig. 9. In relation to depth cues, Fig. 10 depicts the gradations of shadow from multiple light sources [5].

Illustration of Fluids

While the mathematical description of fluids were yet to be described by the Navier-Stokes partial differential equations, [6] turbulent flow was studied by Leonardo da Vinci in numerous visualizations. See Figs. 11 and 12. These illustrations are remarkable because of the dynamic nature of evolving eddies and vortices [7]. This treatment has since been elaborated upon in later centuries (Chaps. 4 and 5).

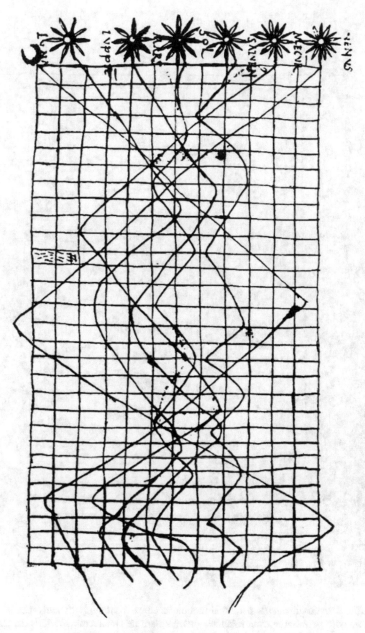

Fig. 1 Reproduction of the first known graphical representation of variables, here the plots of the inclinations of planetary orbits in terms of time. Reproduced with permission from [A Note on a Tenth Century Graph, The Saint Catherine Press Ltd., 1936 Public Domain]

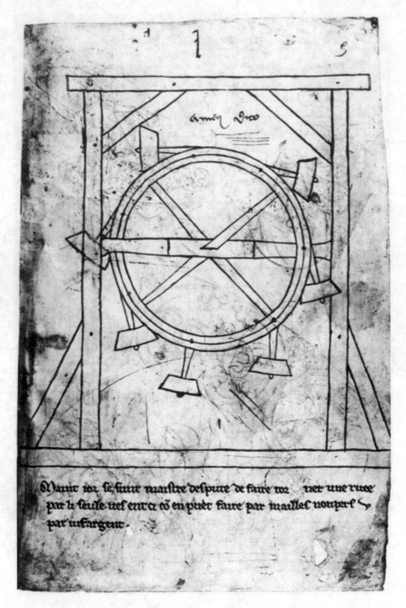

Fig. 2 Villard de Honnecourt's perpetual motion machine from approximately 1225–1235 CE. Reproduced with permission from [Sketchbook of Villard de Honnecourt, Wikimedia Commons, 2005 Public Domain]

Fig. 3 Recreation of the original Gutenburg press at the *International Printing Museum* (Carson, CA). Reproduced with permission from [https://commons.wikimedia.org/wiki/File:PrintMus_038. jpg by vlasta2, Wikimedia Commons, 2011 Creative Commons Attribution 2.0 Generic]

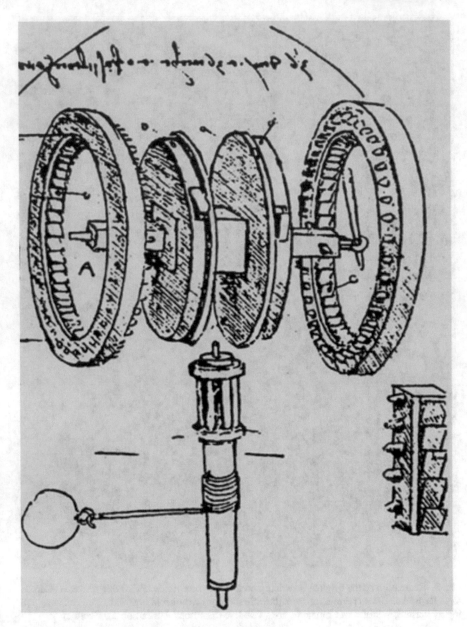

Fig. 4 Leonardo da Vinci (1452–1519 CE). An exploded view of a mechanism. The visualization separates each piece along an axis, including the driving weight. Reproduced with permission from [15th-Century manuscript by Leonardo da Vinci, Wikimedia Commons, Public Domain]

Fig. 5 Cesare Cesariano's (1521 CE) illustrated version of *De Architectura* by the Roman architect Vitruvius (80–15 BCE). The figure is drawn with background grid for length comparisons. Reproduced with permission from [16th-Century manuscript by Cesare Cesariano, Wikimedia Commons, Public Domain]

Analytic Geometry

Illustration in science overlapped progress in mathematics, especially the formative works in analytical geometry. The period of scholasticism (middle ages) differed from ancient Greek development by a deeper concern in the application of mathematics to physical and scientific problems [8]. To clarify, Boyer states,

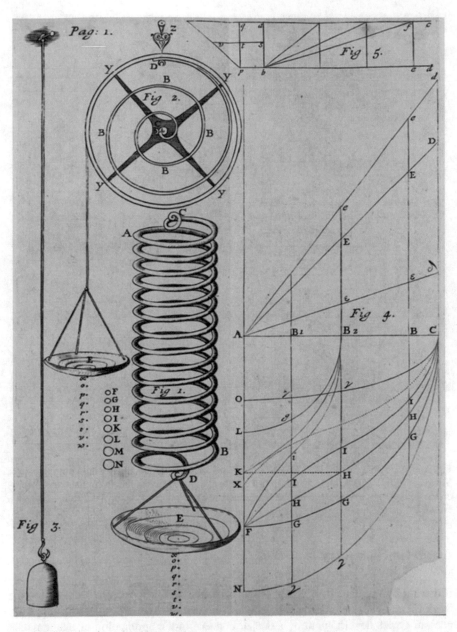

Fig. 6 Robert Hooke's (1635–1703 CE) theory of deformation measurements in graph form. Reproduced with permission from [17th Century manuscript by Robert Hooke, Wikimedia Commons, Public Domain]

Fig. 7 Leonardo da Vinci's use of cutaway and scale to display the cranial cavity. Reproduced with permission from [15th-Century manuscript by Leonardo da Vinci, Wikimedia Commons, Public Domain]

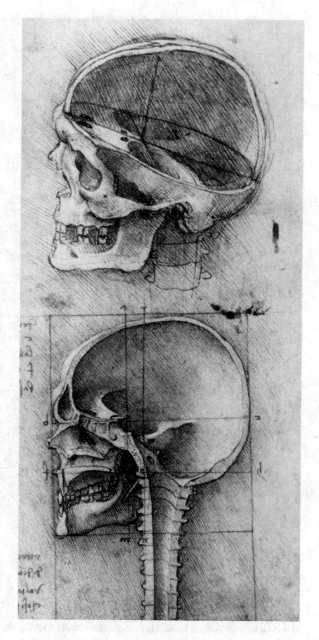

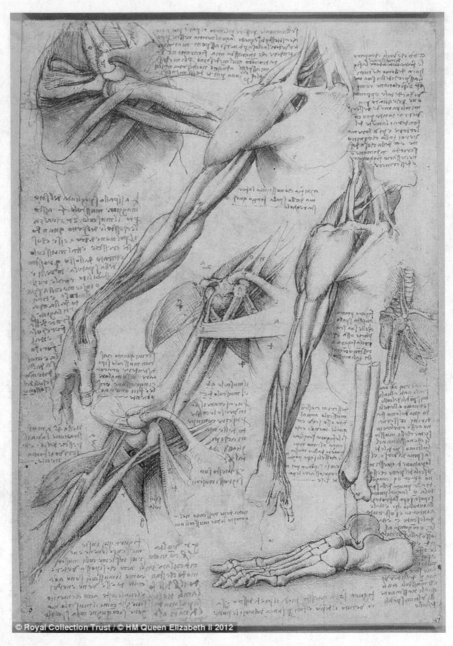

Fig. 8 Transparency use to show musculature in human shoulder and arm. Reproduced with permission from [15th-Century manuscript by Leonardo da Vinci, Wikimedia Commons, Public Domain]

Fig. 9 Perspective projection geometry illustrated by Gaspar Schott. Reproduced with permission from [17th-Century Manuscript by Gaspar Schott, Wikimedia Commons, Public Domain]

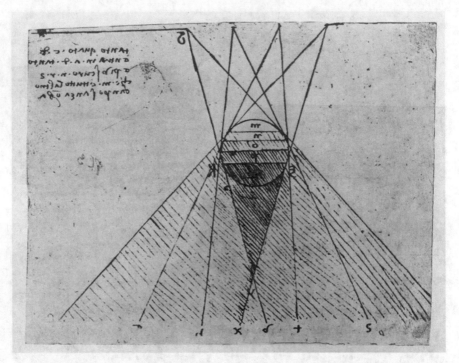

Fig. 10 A study of lighting on a spherical object by Leonardo da Vinci. Reproduced with permission from [15th-Century manuscript by Leonardo da Vinci, Wikimedia Commons, Public Domain]

Fig. 11 Da Vinci's turbulent flow visualization. Reproduced with permission from [15th-Century manuscript by Leonardo da Vinci, Wikimedia Commons, Public Domain]

Fig. 12 Da Vinci's depiction of flow around a barrier with free-surface movement. Reproduced with permission from [15th-Century manuscript by Leonardo da Vinci, Wikimedia Commons, Public Domain]

> The Greeks had built up an elaborate mathematical theory, but they had applied only the most elementary portion of it to science; the scholastic philosophers of the 14th century, on the other hand, possessed the most elementary quantitative study tools but sought ambitiously to make an elaborate quantitative study of science. [2]

The medieval conception of "latitude of forms," therefore, was the precursor to algebraic (analytic) geometry and its application to physical variation. The terminology of the latitude of forms, i.e., "intensio" and "remissio," referred to the change in physical quantities with respect to time.

With this ideation, Nicole Oresme (1323–1382 CE) first applied latitude of forms to geometric representation. Prior to this, Greek mathematical doctrine dictated numbers as discrete, whereas geometrical magnitude continuous—precluding graphical (geometric) depiction of the quantities of physical measurement. However, in Oresme's treatise *Tractatus de Latitudinous Formarum*, he states measured quantities (but not necessary discrete number) could be graphically drawn as points, curves (lines predominated his work), and surfaces [9].

The application of coordinates for the representation of algebraic equations (higher order nonlinear curves) did not begin until Pierre de Fermat (1607–1664 CE) and René Descartes (1596–1650 CE). Their own work relied on contributions from Francois Viéte (1540–1603 CE), whose use of vowels (letters) for variables (parameters) in equations was needed first as a prior addition to algebra.

In Fermat's *Ad Locos Planos et Solidos Isagoge*, the concept of a variable as a geometric extension to be visualized differed from Viéte, whose algebraic variables represented a fixed value in relation to an equation (when all other parameters were themselves fixed). Before Fermat, indeterminate equations of two variables did not have specified meaning. In the words of Fermat,

> Whenever in a final equation two unknown quantities are found, we have a locus, the extremity of one of these describing a line, straight or curved.

Fermat thus began the study of curves defined by equations.

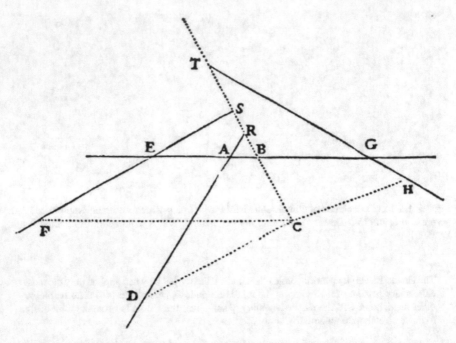

Fig. 13 Decartes graphical depiction of producing points for a straight line as described by the Pappus problem with four lines. One can see the dotted lines (points, or sometimes called loci) are those used in the construction of the curve (line) satisfying the required conditions. Reproduced with permission from [17th-Century Manuscript by René Decartes, Open Court Publishing Co., 1925 Public Domain]

For the most part, Decartes contribution to analytic geometry was the inverse of Fermat's. He produced equations from points [10]. He also introduced modern notation of the algebraic variable and powers. Decartes was not the first to introduce a coordinate system for plotting points, but supported a shift in conceptualization of the relationship between curves and their equations, by providing the methodology for doing so. In his words,

> All geometrical problems may be easily reduced to such terms that afterwards one only needs to know the lengths of certain straight lines in order to construct them.

Coordinate geometry would be necessary both for modern use of mathematics in science and also for the representation of surfaces for rendering by a computer, as seen in Chap. 4 (Fig. 13).

Descriptive Geometry

Descriptive and analytic geometry differ. On the one hand, descriptive geometry is used chiefly to draw (visualize) objects for design and construction, or scientific application. By contrast, analytic geometry is primarily the application of algebra for delineation of curves (straight lines, etc.).

Descriptive geometry is useful for length comparison, when distance information is ambiguous due to foreshortening from perspective projection (i.e., normal viewing through the eye or camera). For example, Fig. 14 shows an orthographic projection contrasted with perspective drawing. This type of projection is a product of descriptive geometry.

Earlier works by Albrecht Dörer and Guarino Guarini, in 1525 and 1665, respectively, presaged the arrival of descriptive geometry. However, Gaspard Monge (1746–1818) was the first to mathematically formalize descriptive geometry in 1794, with his treatise entitled, *Geometrie Descriptive* [4]. Before *Geometrie Descriptive*, Monge served as a draftsman for the French military. He was responsible for the conception of Earth Mover's Distance (EMD)[1] in 1781–prior to the advent of descriptive geometry.

An anecdote from Monge's career, not surprisingly, was that during his military service, he was required to optimize a fortification. This practice was usually performed using laborious calculations. Forgoing the traditional approach, Monge was able to reduce considerably the time required for the optimization task using visualization methods from descriptive geometry [11].

Mechanized Computation

Blaise Pascal (1623–1662 CE) invented the first successful mechanical calculator, today called the Pascaline. It was designed in 1642 and first produced in 1645.

The machine performs addition and subtraction only. Multiplication and division are implemented by repeated addition or subtraction operations. The Pascaline's carry mechanism made it a unique advancement in mechanical calculation. It also featured lantern gears, which up to that time had only been used in much larger water wheels and tower clocks. (See Fig. 15).

The device was initially created to ease the burden of calculations performed in tax collection. However, its use has lead to modern computing, from which today's scientific visualization depends.

---

[1] EMD is a distance metric for measuring the difference between two or more probability distributions. Some of its uses today are measuring the difference between images in computer graphics, or data in scientific visualization.

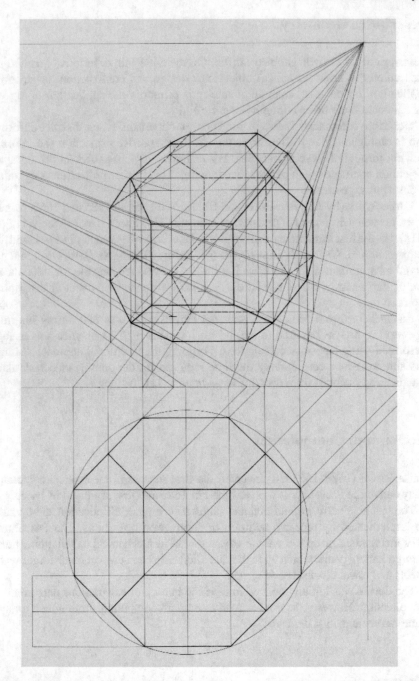

Fig. 14 Perspective and orthographic projections of a polyhedron. In this example, a two-point perspective drawing is shown above the orthographic projection, from the ground plane. Orthographic projection assists visualization by retaining original object length in the direction of rays, thus aiding visual analysis via multiple views. From [Luciano Testoni, Wikimedia Commons]

Fig. 15 The Pascaline, the first functional mechanical calculating machine (originally constructed in 1654 CE). This view displays the row of dials for input. From [Scientific instruments at the Musée des Arts et Métiers by Rama, Wikimedia Commons]

Summary

In the fifteenth century, the scholastics began to revive ancient Greek geometry. However, a greater emphasis was placed on the application of mathematics to physical phenomena.

The printing press allowed the introduction of the treatise. Treatises were the first to feature technical and scientific illustration. Because of printing, the illustration of Leonardo da Vinci and other prominent visualization efforts spread throughout Europe. Some of the unique characteristics of these renderings were cutaways, transparency, and realistic shading.

The systematic association of curves with algebraic equations—of more than one unknown variable—was introduced in the seventeenth century. Fermat and Decartes offered inverse, but equivalent, depictions of analytical geometry. This period also saw the first functional mechanical calculating machine, known today as the Pascaline.

References

1. Robertson, J.: The Enlightenment: A Very Short Introduction. OUP Oxford (2015)

2. Boyer, C.B.: History of Analytic Geometry. Courier Corporation (2012)
3. Funkhouser, H.G.: A note on a tenth century graph. Osiris **1**, 260–262 (1936)
4. Rovida, E.: Machines and Signs: A History of the Drawing of Machines, vol. 17. Springer Science & Business Media (2012)
5. Richter, J.P. et al.: The Notebooks of Leonardo da Vinci, vol. 2. Courier Corporation (1970)
6. Wolfram, S.: A New Kind of Science, vol. 5. Wolfram Media Champaign, IL (2002)
7. Smith, P.H.: Artists as scientists: nature and realism in early modern Europe. Endeavour **24**(1), 13–21 (2000)
8. Curley, R. et al.: Scientists and Inventors of the Renaissance. Britannica Educational Publishing (2012)
9. Funkhouser, H.G.: Historical development of the graphical representation of statistical data. Osiris **3**, 269–404 (1937)
10. Descartes, R.: The Geometry of René Descartes: With a Facsimile of the First Edition. Courier Corporation (2012)
11. Macqueen, W.M.: Solid geometry and industrial drawing. Education+ Training (1959)

Scientific Visualization in the Nineteenth Century

Abstract The scientific illustration of the Enlightenment was largely representational rather than speculative. It conveyed direct observation by realistic representation. There were exceptions, e.g., Decartes fanciful visualizations of magnetic forces. However, this contrasts the scientific visualization of the nineteenth century and later, which became more abstract and speculative, while grounded in empiricism. From the conceptualization of the electric and magnetic field lines of Faraday to the elucidation of molecular structure, visualization was becoming an extension of observation, used to infer scientific knowledge.

The Electromagnetic Field

The electromagnetic field concept represents one of the greatest achievements of the nineteenth century. Its history is rife with the application of scientific visualization. For instance, Descartes' proposed "spiral effluvia" in the seventeenth century as an explanation of magnetism, shown in Fig. 1. This view was tied to much earlier observations of magnetized metals aligning with the earth's own magnetic field (Fig. 2).

In 1820, Han Christian Ørsted observed compass needle deflection when current is either first applied or removed from a circuit. (The compass used in this historical achievement is shown in Fig. 3.) Ørsted's discovery was the first realization that electricity is connected with magnetism.

Faraday's Visual Methods

Not until Michael Faraday (1791–1867) experimentally derived visualizations of electromagnetic phenomena, [1] was there sufficient data to form a proper theory, i.e., James Clerk Maxwell's (1831–1879) equations of electromagnetism. Faraday's visual methodology was to aid in the discovery of electromagnetic induction, that is the creation of current from a moving magnetic field. He also mapped out the

© The Author(s), under exclusive license to Springer Nature Switzerland AG 2022 39
B. E. Hollister and A. Pang, *A Concise Introduction to Scientific Visualization*,
https://doi.org/10.1007/978-3-030-86419-4_3

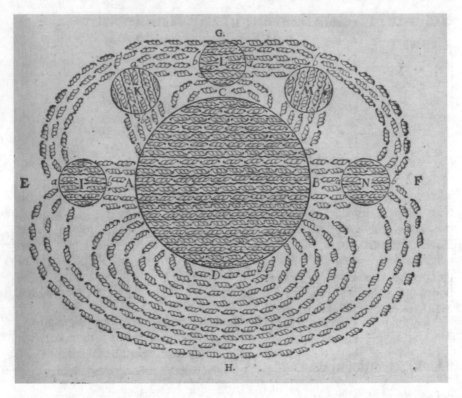

Fig. 1 Decartes spiral effluvia as a visual explanation for magnetism. Reproduced with permission from [17th-Century Manuscript by René Decartes, Wikimedia Commons, Public Domain]

spatial orientations of electromagnetic field lines for numerous arrangements and geometries (Fig. 4). We will see in Chap. 4, that these field lines are more generally considered streamlines, which are used in vector field visualization.

Faraday was not mathematically trained, but rather used his ability for experimentation. The following is Maxwell's own depiction [2] of Faraday's use of visualization from his 1856 paper, [3] titled *On Faraday's Lines of Force*:

> As I proceeded with the study of Faraday, I perceived that his method of conceiving the phenomena was also a mathematical one, though not exhibited in the conventional form of mathematical symbols. I also found that these methods were capable of being expressed in the ordinary mathematical forms, and thus compared with those of the professed mathematicians.
>
> For instance, Faraday, in his mind's eye, saw lines of force traversing all space. Where the mathematicians saw centers of force attracting at a distance, Faraday saw a medium where they saw nothing but distance. Faraday sought the seat of the phenomena in real actions going into the medium, they were satisfied that they had found it in a power of action at a distance impressed on the electric fluids.
>
> When I had translated what I had considered to be Faraday's ideas into a mathematical form, I found that in general the results of the two methods coincided, so that the same phenomena were accounted for, and the same laws of action deduced by both methods, but

Fig. 2 Model of Han dynasty compass circa second-century BCE. Used initially as a fortune-telling device. From [Model Si Nan of Han Dynasty, Wikimedia Commons]

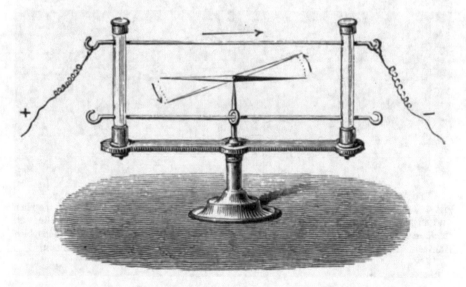

Fig. 3 Ørsted's compass needle. From [Agustin Privat-Deschanel, Wikimedia Commons]

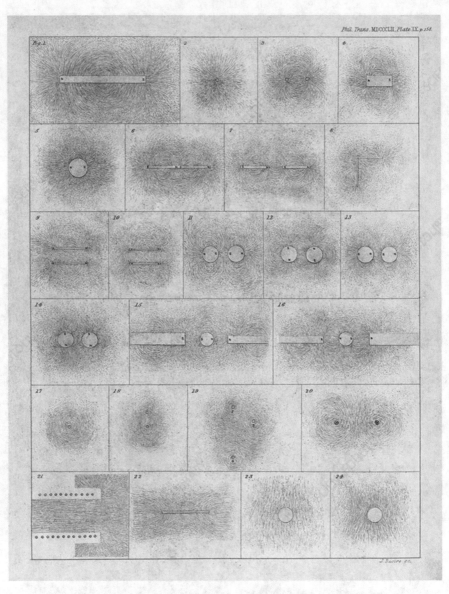

Fig. 4 1852 magnetic field drawings by Michael Faraday. The images are obtained in the plane with alignment of iron filings along the magnetic force. In essence, magnetized iron filings are distributed compasses (or abstractly, vectors). Reproduced with permission from [Philosophical Transactions of the Royal Society, Royal Institution Of Great Britain, 1852 Public Domain]

that Faraday's methods resembled those in which we begin with the whole and arrive at the parts by analysis, while the ordinary mathematical methods were founded on the principle of beginnings with the parts and building up the whole by synthesis.

I also found that many of the most fertile methods of research discovered by the mathematicians could be expressed much better in terms of ideas derived from Faraday than in their original form.

Another example of Faraday's visual methodology, was his depiction of the orthogonality between current (charge) movement, magnetic force direction, and the magnetic field's force upon moving charge (Fig. 5). Faraday's notes accompanying his visualization were the following:

The mutual relation of electricity, magnetism and motion may be represented by three lines at right angles to each other, any one of which may represent any one of these points and the other two lines the other points. Then if electricity be determined in one line and motion in another, magnetism will be developed in the third; or if electricity be determined in one line and magnetism in another, motion will occur in the third. Or if magnetism be determined first then motion will produce electricity or electricity motion. Or if motion be the first point determined, Magnetism will evolve electricity or electricity magnetism. [4]

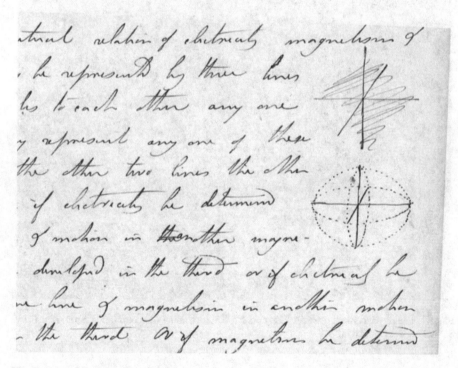

Fig. 5 A page from Faradays' notebook written on 26 March 1832. Reproduced with permission from [19th-Century Notebook of Michael Faraday, Royal Institution Of Great Britain, 1832 Public Domain]

Vector Analysis

Vector analysis is now used extensively in scientific visualization and computer graphics. It should be realized that modern vector analysis did not come about until the latter part of the nineteenth century, [5] after the acceptance of the Gibbs-Heaviside system which itself was derived from the multidimensional complex number system called quaternions.

One may speculate about its earliest visual influences: weather vanes, compass needles, or earlier natural predecessors—e.g., fields of wild grass blown by wind currents (Fig. 6) [6]. Each of these devices allow for the visualization of otherwise invisible movement (air fluid) or force potential, i.e., a compass needle placed in a magnetic field. Fig. 7 shows the oldest surviving weather vane from 820 CE. *Huainanzi*, a Chinese work written approximately 140 BCE, is the first known depiction of a weather vane, referred to as a "wind-observing fan," or *hou feng shan*.

Fig. 6 Field of wind swept grass. A likely conceptual forerunner of vector fields. From [Dustin V. S., Wikimedia Commons]

Fig. 7 Oldest extant weather vane, the Gallo di Ramperto. Photo by Wolfgang Moroder. From [Gilded Weather Vane Signed M..oaldo AD 820 in the Museo di Santa Giulia complex in Brescia, Wikimedia Commons]

Force Parallelograms, Quaternions, and the Gibbs-Heaviside System

Before the formal introduction of vectors as mathematical objects (with arithmetic-like operations), there were parallelograms of forces. Earlier still were parallelograms of velocities. Such accounts of physical quantities were, to some extent, available since ancient Greece for velocity. Force parallelograms were introduced later in the sixteenth and seventeenth centuries [7, 8].

Implicit in parallelogram visualizations of force (or velocity) is the notion of vector addition and subtraction (Fig. 8). Later developments led to the geometric representation of complex numbers, where the real part is graphed along the horizontal axis, and the imaginary part along the vertical axis of the Cartesian-coordinate system (recall the section entitled *Analytic Geometry* from Chap. 2). This two-dimensional representation of complex numbers was first presented by Caspar Wessel (1745–1818) in 1797 [5].

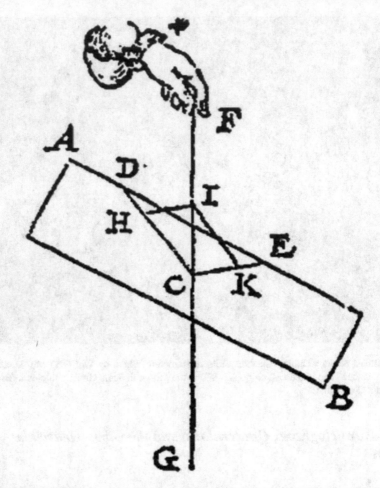

Fig. 8 Representation of a weight supported by a single string, or combination of two strings. From *Les Oeuvres Mathématiques de Simon Stevin* (1634). Reproduced with permission from [Les Oeuvres Mathématiques de Simon Stevin, 17th-Century Treatise, 1634 Public Domain]

Along with the geometric representation of complex numbers and parallelograms of velocities or forces, Gottfried Leibniz (1646–1716) is quoted in a letter to Christian Huygens from 1679,

> I am still not satisfied with algebra, because it does not give the shortest methods or the most beautiful constructions in geometry. This is why I believe that, so far as geometry is concerned, we need still another analysis which is distinctly geometrical or linear...

His proposal did not become our present one, but an influence on vector analysis.

Quaternions were created by William Rowan Hamilton (1805–1865) in 1843. Once again, quaternions, like Leibniz's proposal for such a system, did not become our present-day vectorial system—although they are still used in computer graphics and other fields. They have common elements with vector analysis, however.

Firstly, Hamiliton developed quaternions in an attempt find a three-dimensional system for complex numbers. A quaternion can be expressed as a four-tuple. This is an ordered set of four real numbers (a, b, c, d), three of which are multiplied by the imaginary numbers i, j, and k.

The general form of a quaternion is $a + bi + cj + dk$. Operators for quaternions maintain most properties associated with real numbers operators, except for commutation when multiplying. Note that in modern vector analysis, the cross product of vectors does not conserve either the associative nor commutative properties.

We have already seen the influence of scientific visualization (Faraday's work) on electromagnetic theory. Maxwell had originally expressed his equations in a longer format, and explicitly for each spatial dimension. In later publications, he extended the notation to use Hamilton's quaternion. However, this new formulation was still ungainly, and thus preventing further wide-spread acceptance [9].

Around 1880, Josiah Gibbs (1839–1903), and Oliver Heaviside (1850–1925), independent of each other, set about to simplify the electromagnetic field equations created by Maxwell. In the creation of the modern vectorial system, Gibbs and Heaviside extracted the imaginary part of the quaternion, the three-dimensional geometric component which was needed. This part of the quaternion was repurposed as the modern vector, along with Heaviside' addition of the ∇ operator[1] (among other modifications).

Interestingly, ∇ (referred to as "nabla") bears similarity to Hamilton's ⊳ operator. However, it is visibly rotated. Some might say it is made, "right side up," especially those that had modernized quaternions to form vector analysis. The word "nabla" is also the Hellenistic Greek word for a Phoencian harp (Fig. 9), due to its similar appearance. (Note our earlier discussion of glyphs from Chap. 1.)

Crystallography

Another nineteenth-century discipline that used visualization extensively was the study of crystallography. This field began when René-Just Haüy (1743–1822) undertook the systematic investigation of calcite (Fig. 10) [10]. Crystalline solids of elements or compounds are regularly repeating structures expressed as a macroscopic polyhedron. However, before the discovery of X-rays and diffraction from crystals, scientists were concerned with their observable polyhedral forms, a central puzzle being the proper way to characterize them.

While Haüy had been a proponent of a theory emphasizing that all members of a species of crystals shared the same angles between crystal facets (the faces of their polyhedral shapes), Christian Samuel Weiss (1780–1856) in 1817 set upon a more systematic characterization. He created the notion of crystal "zones" [11]. This

---

[1] $\nabla = \left(\frac{\partial}{\partial x}, \frac{\partial}{\partial y}, \frac{\partial}{\partial z} \right) = i \frac{\partial}{\partial x} + j \frac{\partial}{\partial y} + k \frac{\partial}{\partial z}$

Fig. 9 Harp. From
[Webster's Dictionary,
Wikimedia Commons]

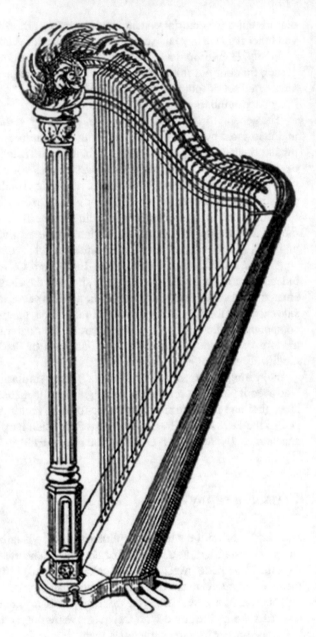

Fig. 10 Calcite crystals showing individual facet. From [Joan Rosell, Wikimedia Commons]

approach proved fruitful, as Haüy's method did not offer a consistent agreement among crystal species.

A crystal "zone" as defined by Weiss, is the set of planes all mutually perpendicular to the same axis, i.e., the "zone axis." The utility in this, is that by recording precisely the entirety of the zones for a crystal type, it becomes possible to determine its species regardless of small irregularities between specimens. Additionally, from a knowledge of a crystals zones, the relative angles between intersecting planes can be determined.

Franz Ernst Neumann (1798–1895), Weiss' student, created a visualization system to view the entirety of a crystal's zones and their intersections, including where zones intersect forming triangular facets. Neumann's stereographic-projection visualization influenced later visualization for the depiction of rotational symmetries (i.e., the Gadolin 'stereographic' projection). Neumann's stereographic method was published in 1823, and is still used today.

The projection of each facet is shown in Fig. 11. The visualization can be seen in Fig. 12. The small circles depict each of the facets projected onto the sphere in a single view. As facets border other facets from different zones, each small circle occupies more than one elliptical trace outlining the configuration space of a particular zone of parallel planes in the crystal.

The value of Neumann's visualization (as with many modern visual methods) is to provide a summary view not available by direct observation alone. A visual summary enables the user to make conclusions about the test data or object, in this instance, a crystal specimen. Here, crystal categorization can be derived from Neumann's method.

Visualization in the Birth of Structural Chemistry

Before Friedrich August Kekulé (1829–1896), there was no firm foundation for the internal structure of chemical species. While work had been done by James Dalton (1766–1844) on his atomic theory (Fig. 13) to explain gasses, chemical reagents were considered beyond the realm of geometry [12].

Beginning with the study of carbon-containing chemical compounds, it had been observed through empirical formulations (i.e., those derived by chemical decomposition) that consistent ratios of elements occurred in a given chemical species (compound). In these empirical combinations of constituent elements, the theory of valency, i.e., chemical bonding, came about.

Kekulé suspected that valency, that is the number of bonds formed by an element, was mostly constant. For instance, carbon consistently forms bonds with four other elements of various types, at other times two. Using the idea of fixed valency, he first proposed linear structures of carbon as the "nucleus" of multiple chemical structures. His contemporary, Archibald Scott Couper (1831–1892), was the first to represent these carbon-based chemical species with line drawings showing the geometric arrangements [13].

Most notably, Kekulé proposed the structure of benzene (a compound discovered by Faraday) in 1865. (See Fig. 14.) Its theoretical success was in predicting then unknown isomers, species with the same ratios of elements but varying in chemical properties and thus structural linkage.

Some have questioned the veracity of Kekulé account of benzene's structural elucidation [14]. However, the form of visualization in his account was that of the superposition of imagery (a psychological state called homospatial processing). He envisioned the carbon atoms (originally in a linear linkage) where the ends became

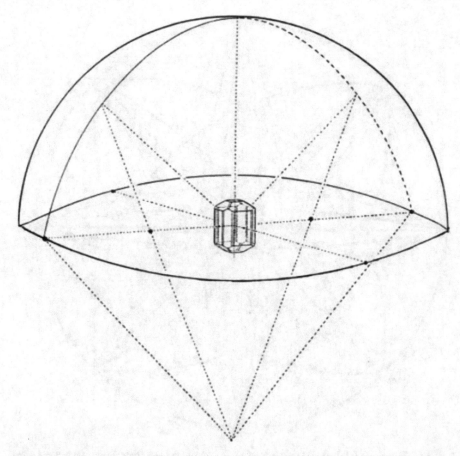

Fig. 11 A perpendicular ray is projected to intersect with a sphere centered on the crystal, for each facet. Then for each point on the top hemisphere, a line is connected to the south pole of the sphere and intersected with the equitorial plane. From [Crystallography from Haüy to Laue: Controversies on the Molecular and Atomistic Nature of Solids, IUCr Journals, 2012 Open Access]

connected, along with the imagery of a snake eating its own tail. In Kekulé's own words:

> Again atoms fluttered before my eyes. Smaller groups stayed mostly in the background this time. My mind's eye, sharpened by repeated visions of this sort, now distinguished larger figures in manifold shapes. Long rows, frequently linked more densely; everything in motion, winding and turning like snakes. And lo, what was that? One of the snakes grabbed its own tail and the image whirled mockingly before my eyes.

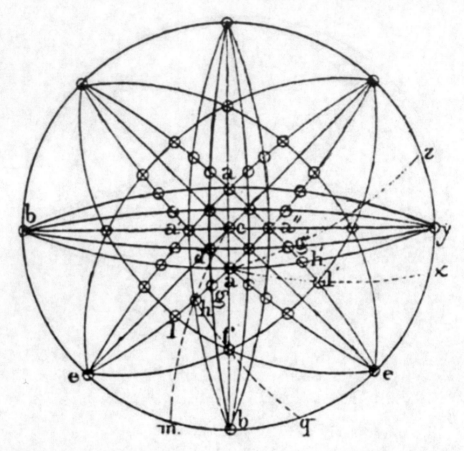

Fig. 12 Neumann's stereographic projection of crystal planes and their facets. Each facet edge between zones is labeled, where the circles mark particular crystal faces (from Kubbinga). From [Crystallography from Haüy to Laue: Controversies on the Molecular and Atomistic Nature of Solids, IUCr Journals, 2012 Open Access]

Programmable Mechanical Computation

Charles Babbage (1791–1871) created two forerunners to the modern electronic computer. In 1822, he designed the Difference Engine (Fig. 15) to employ numerical methods of finite differences to solve polynomial functions (see Chap. 2 on Analytical Geometry).

Later, Babbage was to design the first programmable mechanical computer called the Analytical Engine which allowed for memory and branching instructions, making it so-called Turing-Complete. However, Alan Turing's (1912–1954) theoretical work on computation was not to done until the twentieth century.

Neither the Difference Engine or the Analytical Engine were built during Babbage's lifetime [15].

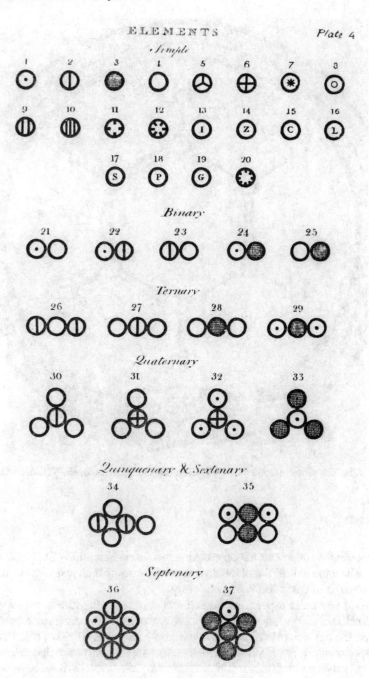

Fig. 13 Dalton's visual models of atoms. From [Plate IV of John Dalton's "A New System of Chemical Philosophy," Wikimedia Commons]

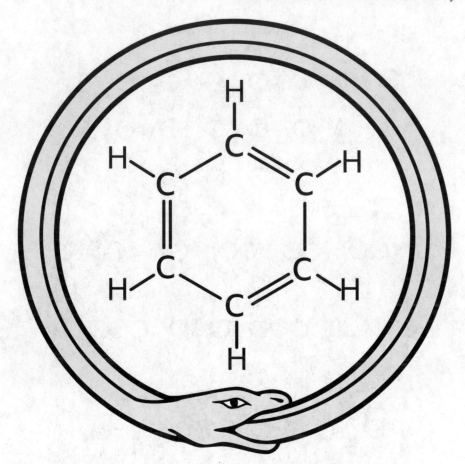

Fig. 14 Structure of benzene and the August Kekulé story. From [Haltopub, Wikimedia Commons]

Summary

The nineteenth century saw a sophistication in science. Scientific visualization during this time became an additional tool in its own right, more than just a means to convey direct observation as it had been previously.

Beyond any other aspect, the unseen world was coming into view using scientific visualization. We have seen only some examples of the nineteenth century. For instance, the physicist Henri Poincaré was known to work visually [16]. There were also experimentally based visualizations such as the Chladni plates that help visualize contours between vibrating sections on surfaces. Chladni plates have been suggested to have had an influence on Faraday's own visualizations of the electromagnetic field [17]. However, the value of scientific visualization at this time is no more apparent

Fig. 15 A functional difference engine on display at the Computer History Museum in San Jose, CA. From [Allan J. Cronin, Wikimedia Commons]

than in the elucidation of the field concept in physics, and theories of the structure of matter using only deduction from chemical reactions.

Computing continued its advancement with Babbage's analytical machine, although practical computing was still to come about only in the next century (Chap. 4) when a confluence of both computing and scientific visualization would take place.

References

1. James, F.A.J.L.: Michael Faraday: A Very Short Introduction, vol. 253. Oxford University Press (2010)
2. Harman, P.M.: Maxwell and Faraday. Euro. J. Phys. **14**(4), 148 (1993)
3. Maxwell, J.C.: On Faraday's lines of force. Trans. Camb. Philos. Soc. **10**, 27 (1864)
4. Al-Khalili, J.: The birth of the electric machines: a commentary on Faraday (1832) 'Experimental Researches in Electricity'. Philos. Trans. R. Soc. A: Math., Phys. Eng. Sci. **373**(2039), 20140208 (2015)

5. Crowe, M.J.: A History of Vector Analysis: The Evolution of the Idea of a Vectorial System. Courier Corporation (1994) 0
6. Bell, G.D.H.: The History of Wheat Cultivation. Springer, Berlin (1987)
7. Miller, D.M.: The parallelogram rule from pseudo-aristotle to newton. Arch. Hist. Exact Sci. **71**(2), 157–191 (2017)
8. Blockley, D.: Structural Engineering: A Very Short Introduction. OUP Oxford (2014)
9. Forbes, N., Mahon, B.: Faraday, Maxwell, and the Electromagnetic Field: How Two Men Revolutionized Physics. Prometheus Books (2014)
10. Glazer, A.M.: Crystallography: A Very Short Introduction, vol. 469. Oxford University Press (2016)
11. Kubbinga, H.: Crystallography from Haüy to laue: controversies on the molecular and atomistic nature of solids. Acta Crystallogr. Sect. A: Found. Crystallogr. **68**(1), 3–29 (2012)
12. Brock, W.H.: Norton History of Chemistry. WW Norton (1993)
13. Hepler-Smith, E.: Just as the structural formula does: names, diagrams, and the structure of organic chemistry at the 1892 Geneva nomenclature congress. Ambix **62**(1), 1–28 (2015)
14. Rothenberg, A.: Creative cognitive processes in Kekule's discovery of the structure of the benzene molecule. Am. J. Psychol. 419–438 (1995)
15. Hutton, D.M.: The difference engine: charles babbage and the quest to build the first computer. Kybernetes (2002)
16. Cumo, C.: Henri Poincaré: a scientific biography, by Jeremy Gray. Can. J. Hist. **48**(3), 503–505 (2013)
17. Comer, J.R., Shepard, M.J., Henriksen, P.N., Ramsier, R.D.: Chladni plates revisited. Am. J. Phys. **72**(10), 1345–1346 (2004)

A Convergence with Computer Science

Abstract Prior to the latter part of the twentieth century, the field of scientific visualization *was* both technical illustration [1, 2] and analytic geometry [3] when used to depict scientific enterprise. It was not directly involved in the emergence of computational theory [4]. However, computers became an essential tool for *interactive* scientific visualization, via the use of the vacuum tube display.

Government involvement in the initial development of computers centered around scientific computation in military and defense applications. These areas constituted much of the work of John von Neumann, both at Los Alamos and the *Institute for Advanced Studies* in the United States. Early computers were the workhorse for numerical simulations in ballistics, atomic physics, and later weather simulation [5]. Thus, the emerging field of computer science was primed for its eventual confluence with scientific visualization.

Von Neumann's Architecture

The modern organization of computing systems owes much of its design to a group of engineers and scientists at Princeton University. Their achievement was that of the *stored-program* concept (Fig. 1), implemented first in the *Electronic Discrete Variable Automatic Computer* or EDVAC for short [6]. The EDVAC was constructed in late 1944 at the Moore School of Electrical Engineering.

Before EDVAC, electronic computational systems such as the ENIAC [7] had to be reconfigured for each separate type of computation [8]. While Charles Babbage's earlier mechanical *Analytical Engine* was conceived of as programmable with conditional logic [9], it was unrealized fully during the nineteenth century due to numerous practical issues.

The design for the EDVAC was laid out first in a report by John von Neumann. The "First Draft of a Report on the EDVAC" was hand-written by von Neumann while traveling by train from Princeton to Los Alamos, where he was involved in the *Manhattan Project* for the development of nuclear weaponry.

© The Author(s), under exclusive license to Springer Nature Switzerland AG 2022   57
B. E. Hollister and A. Pang, *A Concise Introduction to Scientific Visualization*,
https://doi.org/10.1007/978-3-030-86419-4_4

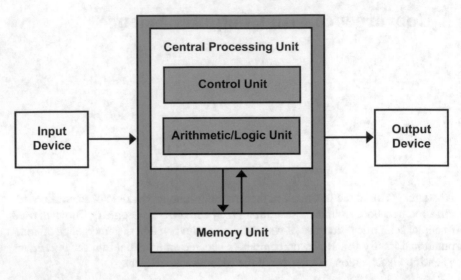

Fig. 1 The "von Neumann" architecture. The stored program was kept in the memory shown in upper-left and upper-middle units of the diagram. From [Kapooht, Wikimedia Commons]

Funding for the EDVAC was overseen by the US Army Research Laboratory's Ballistics Research Laboratory, and this was to have an influence on its use [10]. Furthermore, the invention of the EDVAC was instrumental in von Neumann's later efforts at Los Alamos, and reflected in such research publications as "A Method for the Numerical Calculation of Hydrodynamic Shocks" [11].

Improved Numerical Calculation

The first programs written for computers performed calculations for scientific application. While the EDVAC was the impetus for "scientific computing," over time, the EDVAC was replaced with more efficient transistor-based machines, such as the TRADIC (TRAnsistorized DIgital Computer) [12].

These days, integrated circuit technology has allowed considerable reduction in both size and power consumption. Consequently, the number of transistors contained in modern computers has vastly increased. It is not uncommon for a typical desktop computer to contain half-a-billion transistors, compared to the very modest number of semiconductor components (less than two thousand) of the original TRADIC. The TRADIC also occupied the size of a small room.

What remained consistent from the outset was the application to scientific problems, of which the electronic computer itself was borne. Although computers of the 1940–1950s are seen as relatively slow by today's standards, their purpose was initially to solve for values used in mathematical models.

Such *numerical analysis* had an even earlier start than that of the computer [13]. However, techniques in this area of mathematics could now operate on a much larger scale, and with shorter time frames, through computer automation. Iterative solutions, characteristic of numerical methods, require many iterations for an accurate approximation. These methods benefit highly by using computer calculation.

An elementary example is Newton's method, first developed by Sir Isaac Newton in "De analysi per aequationes numero terminorum infinitas," and published in 1711. The method approximates the root of a univariate function by solving for the x-intercept of a tangent (i.e., the derivative) at a value sufficiently close to the root. Iterations converge to a solution, where each succession uses the x-intercept solved in the previous pass, to find a new tangent.

Numerical solution requires the translation of mathematical models of physical phenomena to a form that computers can solve numerically (versus analytically), especially for cases where there is no closed-form solution. For physical systems, a common example is the Navier-Stokes equation, using numerical methods [14, 15] that "discretize," or chop into pieces, finite elements of a fluid. The time variable is treated similarly, so the fluid evolves each *time step*, and not continuously.

Computer Graphics and Utah

Early in the study of computer graphics (i.e., CG), widespread classification of further specialization was uncommon. For instance, papers on applying CG to scientific problems were still presented at conferences devoted to CG research of all types, from CG cartoons to molecular modeling. Some of the initial research dealt with interfacing the television (and other peripherals such as a plotter printer [16]) and drawing realistic imagery.

It was at the University of Utah that much of the pioneering research was performed. The university was one of the first to produce algorithms for interactive lighting, while advancing the earlier research efforts of Ivan Sutherland (known as the "father of computer graphics" [17, 18].

Distinguished alumni went on to found companies that have led to developments in the field of, chief among these, *Pixar*, and the most relevant to scientific visualization, *Silicon Graphics*. *Evans and Sutherland*, the very first company, was also first to produce a commercial *frame buffer* (Fig. 2), called the PS340 [19].

A frame buffer stores a *bit-mapped* or per-pixel representation of an image. As we will see, raster displays need to have information about the display for every location on the screen. A "buffer" of memory between the computer and the screen is essential for efficient refresh of the television.

Today, the legacy of research at the University of Utah has transformed, as represented by the *Scientific Computing and Imaging Institute*, into a major research center for visualization.

Fig. 2 In 1974, Evans and Sutherland produced the PS340 frame buffer, as an add-on to their Picture System product. Shown here is a molecular raster image obtained from the PS340. Aerospace and pharmaceutical companies were some of the many corporate consumers of the early technology. Reproduced with permission from [Raster graphics (PS340), Ethan A. Merritt—http://skuld.bmsc. washington.edu/people/merritt/graphics/ps330/ps330.html, 2002]

Of the University of Utah's distinguished alumni, Jim Blinn has been notable for his doctoral research in the fundamental of *texture mapping* [20] and *bump mapping* [21]. Texture mapping, in its basic form, is a method to apply a texture, usually an image, to one or more geometric primitives. It can be as simple as a picture of colors, or may be used as a lookup table for lighting calculations. Bump mapping is a method that attempts to introduce complexity into the scene without additional geometry. It does this by using lighting calculations to simulate a rough surface even though the surface is flat. In both algorithms, low- versus high-resolution geometry can be made to appear detailed (e.g., textured or bumpy). These mappings require less overall computation and storage.

Fig. 3 Voyager 2's encounter with Saturn in 1981. From the official NASA video animation by Jim Blinn and Charles Kohlhase. From [Voyager II Saturn Flyby 1981 NASA Video, US Federal Government, 1986 Public Domain]

While at the *Jet Propulsion Laboratory* in Pasadena, CA, he pursued work that would be considered scientific visualization for the masses. He pioneered the creation of such animations for the television and educational programs as "Cosmos," "The Mechanical Universe," and of the Voyager space probe mission (Fig. 3). These computer animations explained scientific and mathematical concepts in a form that is more accessible to the general public.

Raster Pipeline

The raster pipeline refers to the sequence of operations that transforms representations of data and geometry into images. It is usually broken into several major steps as illustrated in Fig. 4 and implemented in today's hardware. Picture elements, or the aforementioned "pixels," are calculated for each geometric primitive projected from three-dimensional space to device coordinates of the screen. Key to the computer renderings for scientific visualization was contributions to this pipeline.

Research performed by Jim Blinn, and his contemporaries, was that of building stages of the "rendering pipeline." First, determining the most efficient models to

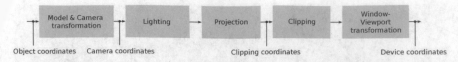

Fig. 4 Raster pipeline. From [Vierge Marie, Wikimedia Commons]

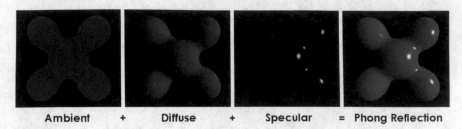

Fig. 5 The Phong reflection model. From [Brad Smith, Wikimedia Commons]

represent geometry, and then algorithms to process the geometry. We will see that the *raster pipeline* (Fig. 4) is itself only a piece of a higher order flow of information.

The raster pipeline made rendering efficient enough to display adequate data to represent images, along with appropriate simulation of light to provide three-dimensional cues. This being done in such a way as to allow users to interact with, or move their viewpoint about, a scene. Another early competing technology was called the "vector" display or pipeline, and was capable of line-drawing only. As dedicated memory for images became less expensive to produce (i.e, the frame buffer), along with an increase in overall computational speed, the raster display became the predominant rendering method.

Basic lighting of surfaces require diffuse and specular light [22] (Fig. 5), which determines color and intensity over a surface. After lighting, texture from image data may be provided as described. While meant initially for realistic imagery, textures may be utilized to represent more abstract information about scene geometry, such as data from scientific computation not otherwise visible.

Silicon Graphics

While computer-generated imagery was developing quickly throughout the 1970s, computational bottlenecks were still preventing the rendering of sufficient data at interactive rates. For CG to be a viable tool for scientific visualization, the movement of the rendering pipeline from software, to hardware, was an important next step.

Another doctoral graduate of the University of Utah, James Clark, perhaps a more controversial figure than Jim Blinn [23], was arguably more influential in the general adoption of computers for graphical applications. While as an Assistant Professor at the University of California at Santa Cruz, he began publishing research aimed at increasing the efficiency of the raster pipeline, the first of which was "Hierarchical Geometric Models for Visible Surface Algorithms" [24]. In this paper, he presented methods to cull geometry that do not contribute to the visible part of a scene, thereby speeding up the pipeline. He later went on to found Silicon Graphics, a company that produced graphical workstations.

The Geometry Engine

After having moved to Stanford University, James Clark published another paper on the design of a dedicated co-processor [25], whose purpose was to offload the geometry stage from the main system processor (or CPU). This "peripheral" was called the *Geometry Engine* and is the forerunner of all dedicated graphics hardware.

Modern-Day Graphics Systems

"Graphics cards" for the personal computer have now supplanted workstations, thus removing Silicon Graphics and similar companies from competition in this market. Such hardware is called a GPU, or *Graphics Processing Unit*, to reflect its prominence in the overall architecture of a modern computer. The functionality of the GPU has become more flexible than the original graphics pipeline. Parallel computation by offloading from the CPU is still the underlying principle. GPUs incorporate many thousands of programmable "shader" units, or parallel processors, using single-instruction multiple-data (SIMD) and *gather-scatter* operations [26]. In the next chapter, we will discuss the GPU's growing usage beyond graphics, for such applications as *machine learning*, used to augment the visualization pipeline.

As of this writing, *interactive ray-tracing*, which computes intersections and lighting for every pixel, is becoming widely available. Ray-tracing, while using a "raster" display, is not the same as the raster pipeline. This is due to the fact that each rendered pixel in ray-tracing is not determined by a *pixel-fill* algorithm that resembles scanning over a geometric primitive. Each ray is best computed independently of every other screen pixel, and highly parallel architectures have enabled this for interactive rates. Additionally, it is possible to mix raster pipeline rendering with ray-tracing, and thus ray-tracing can take part for only some of the rendered image produced by a raster pipeline. Ray-tracing is best for realistic rendering of scenes containing reflections of light on smooth surfaces, producing mirrored reflections from multiple scene elements.

Scientific Visualization Algorithms

Spatial data is measured over two or three physical dimensions, and may change with time. Scientific visualization is often tasked with finding and emphasizing spatial features of interest. Naturally then, methods that show these features can be useful indications in helping us understand the data.

When spatial data is produced from first-principles, it is the result of a forward simulation. Elucidation of first-principles from observed phenomena is called inverse simulation.

Simulation of the "forward" variety marks the development of this era. Numerical solution of systems such as Navier-Stokes, as discussed, is an example of a forward simulation. Before looking at some of the most useful visualization algorithms, we need to briefly discuss how data is captured, processed, and filtered before being rendered and eventually visualized by a user.

The Visualization Pipeline

Earlier, we discussed the *raster* or *rendering pipeline*. The *visualization pipeline* (Fig. 6 [27]) is a general conception, differing from the raster pipeline. Rendering occupies a fixed location at the back end, of the larger visualization pipeline.

At the front of the pipeline, we have our source information. For spatial data, this is discrete measurement, recorded at marked intervals over dimensions of interest. The original data, which is unavailable to the pipeline, is inferred from the samples, as the display device will need to draw intervening information not present at the input.

As an example, our raw data may be the output from a simulation. The simulation might be of a material, such as a carbon fiber strut under load. Finite-element analysis may produce a data set of volumetric cells, containing quantitative information about the internal forces present at a given time step. The resolution of the data is limited by the coarseness of the cells.

After export from a simulation code, the data will be in a format that may not be readily available for analysis, say the interpolation of values between cells. The raw data might need to be copied into a different layout, such as the Hierarchical

Fig. 6 The individual steps in the visualization pipeline. Rendering is a subsystem at the back end of the visualization pipeline

Data Format (HDF5, a common scientific data format) [28] or converted to a format that the analysis software can ingest. Additionally, the data may have missing or invalid values. The inappropriate data will need to be ignored or replaced (possibly by averaging nearby valid data).

After having been initially processed, the data is considered "prepared." At this point in the visualization pipeline, an end user may apply a filter to visualize only specific aspects of the prepared data. If the user is interested in the temperature of the material, other values such as strain can be excluded. Later, when we discuss visualization algorithms in more detail, the filtering stage can depend on the method of visualization as well. For instance, there may be numerical values for density that cannot be easily distinguished visually, when using a technique called volume rendering to discern material components. If this is so, then using meta-data (if present) to filter out a given material with the same approximate density as others might be applied before visualization.

Once we reach the *mapping* stage, we enter into the area where visualization algorithms are used. We will discuss these more thoroughly later. However, mapping, depending on the algorithm, will create a pixel color or geometric primitive from the *focus* (filtered) data. This could be a surface or a color from a *transfer function* (which typically maps data values to color values).

In our simulated composite material example, a user could be interested in visualizing a stress fracture as a three-dimensional surface running through the strut. A set of values corresponding to the fracture would generate (map to) the appropriate geometry, via an appropriate visualization algorithm. This might also be performed as an extraction of a subset of the geometry that represents the fracture locations. The geometry produced in this step is then fed to the rendering subsystem.

Field Data

Field data in scientific visualization takes on a similar meaning as it does in everyday terminology. By the term "field," we simply refer to a region. This region may be one dimensional, as in the case of a straight line or curve. In other words, a form or model whose contents can be specified with a single parameter.

The region may also be *multidimensional*, such as a two-dimensional surface, which is likely the common notion of a field (e.g., think of a plot of land with lines of ownership marked relative to guideposts). A two-dimensional surface must still occupy three-dimensional space, as everything physical must contain our familiar dimensions of width, height, and depth. For instance, the plot of land (field) could be on a hill, so there is a third dimension. Location can be specified by latitude and longitude, where elevation is implicit as long as we remain on the ground (surface).

There are actually two contexts for field: (1) "field" as in implying spatial continuity represented discretely and (2) "field" as in *multi-field* or multivariate data.

Scalar Fields

When we look at physical attributes like temperature, air pressure, or chemical concentration, we have values that are presented with a single *real* number. Without discussing continuity theory, let us rest on intuition [29]. These quantities are indivisible in the "real world." They represent measurements, or magnitudes, and scale accordingly. They are, in fact, called *scalars*.

Tables are kept in computer memory to hold information on locations in our field. These quantities are *discrete*. As computer memory is limited, the number of digits used to represent a number is finite. Because of this, there is limited precision in our values.

Sometimes limited precision also requires us to estimate the values for locations not recorded in our tables. Again, as memory is limited, the number of field locations in our table is also finite. This estimation is called *interpolation*, and there are various mathematical methods used for this purpose.

The distinguishing aspect of *scalar fields* is that spatial locations contain only single values. The question then is, "how best to draw this information?," using algorithms executed by a computer.

For scalar fields, there are many options. One familiar method, taken from cartography (Fig. 7), is called *isoline extraction* [30]. An isoline is a line drawn through a field, where the field contains the same value. The use of isolines assumes a continuously varying field of scalars. If the field were random, then these locations would likely not form a line, but a set of disconnected points.

This method can be extended to surfaces embedded in volumes, and then becomes *isosurface extraction*. Now, whole surfaces represent locations with the same value. Considerations of continuity remain.

Many types of scalar fields use isosurface extraction methods. Computational fluid dynamics (CFD) simulation can generate field data that consists of temperature, pressure, or a variety of other physical variables. Visualizing "zones" of temperature, above or below a given threshold, may be important in determining the effects on various elements in a fluid, or the thermal conductivity of a region's material contents (see Fig. 8) [31].

Figure 9 shows isosurfaces of constant electron density, outlining a molecule's crystal structure. The isosurfaces allow researchers to "fit" known chemical linkages to complete the structure of the protein.

Beyond basic isosurface extraction of electron density, there are many modern sophisticated visualization methods developed especially for molecular visualization in recent years [32]. These have a long lineage, extending back to the earliest "ball-and-stick" molecular models, displayed on some of the first rendering systems (like the previously mentioned Picture System).

Marching Cubes [33] is the name given to a widely used algorithm for isosurface extraction. The algorithm works by checking all known scalar values in a data set against a threshold or *isovalue*. The isovalue is the same across all points on the surface. Once all data points have been measured against the isovalue, they can be

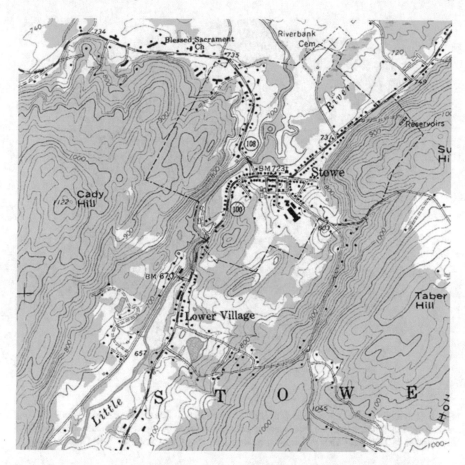

Fig. 7 Topographic map showing contour lines. Lines closer together represent steeper gradations in elevation. From [USGS Digital Raster Graphic, Wikimedia Commons]

assigned as either being inside, or outside, the surface containing values above (or below) the value. Two-dimensional cases are shown in Fig. 10.

There is yet another method called *Volume Rendering* that integrates values from a scalar region [34]. The method does this by sampling values along "rays" running through the volume, for each pixel. This step of gathering samples is called *ray-marching*.

The exact way the algorithm combines the samples depends on what values are intended to be emphasized. Additionally, particular values may be assigned opacity and color in lookup mappings called *transfer functions*. As can be seen in Fig. 11, the bone tissue, which presumably has a different scalar value from muscle, is made more opaque. It occupies the central part of the region to be visualized.

This method excels at allowing users to see into a volume in a single view, instead of having to take subsequent cross sections or isosurfaces to form a full visualization.

Fig. 8 Rayleigh-Bénard convection with Ra = 108 and Pr = 1. Temperature isosurfaces exhibiting ascending hot plumes (red) and descending cold plume (blue). From [Applicability of Taylor's Hypothesis in Thermally Driven Turbulence, Royal Society Open Science, 2018 Open Access]

No intermediate geometry is needed and allows for visualization of the entire data volume. Because of this, the algorithm is often referred to as "direct" volume rendering. The drawback of the method is that it is not always easy to find an appropriate transfer function to properly delineate scalar values. There do exist, however, data and model-driven approaches that seek to optimize transfer function determination [35].

Vector Fields

If we consider data points that consist of more than one value, then we may assign a set of numbers to locations in space. Vectors are collections of scalars, with each scalar being its own component.

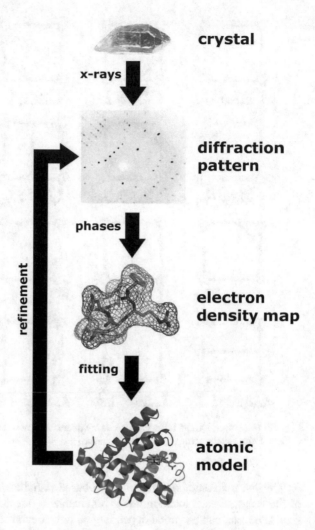

Fig. 9 Isosurfaces derived from an electron density field, before fitting chemical linkages. From [Thomas Splettstoesser, Wikimedia Commons]

The mathematical notation of a vector looks like the following: $(x, y, z)^T$. (The "T" simply stands for *transpose*, so that we can represent a column vector in row format.) Most commonly, vectors are two or three dimensional, as the physical quantities they represent are along a surface or embedded within a volume.

At each storage location in our region, we can record a vector. This vector naturally may represent the velocity of a fluid, potential such as electrical repulsion, or even the gradient (change over region) of a scalar value. The methods applied to such vectors are general, but can be more or less useful for the data sets they are used to visualize.

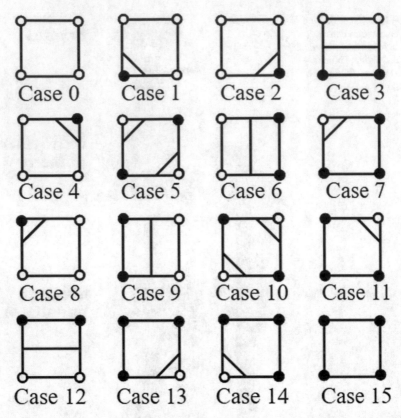

Fig. 10 Isocontour lookup table for Marching Squares (the two-dimensional version of Marching Cubes). From [Nicoguaro, Wikimedia Commons]

The first such method is that of glyph-based visualization [36]. We place a symbol of the vector at each location that is applicable to our field. The choice of location, size, style, etc. can be varied depending on how much information about the vectors a user requests. Figure 12 shows a two-dimensional vector field, with glyphs called *hedgehogs* [37]. These glyphs are also color coded to provide information on the magnitude of each vector.

Another visualization, called *streamlines*, is taken from the particular solution (integral curve) to an ordinary differential equation. Numerical solutions can be calculated using the Euler method, first published in "Institutionum calculi integralis" (1768), which naively adds vector displacements directly but can lead to unstable solutions depending on step size. Or, more accurate and stable approximations can be achieved using such methods as those introduced by Carl Runge and Wilhelm Kutta (i.e., the Runge-Kutta methods) [38].

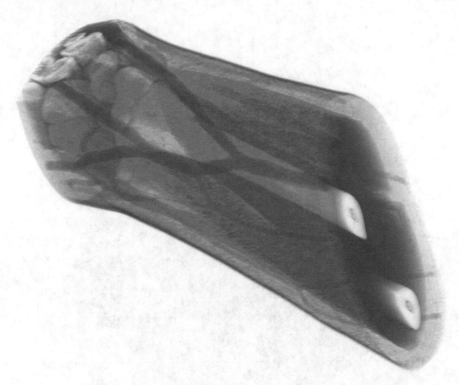

Fig. 11 Volume rendering of the various tissue types in a forearm. From ["Volume rendered CT scan of a forearm with different color schemes for muscle, fat, bone, and blood," by Sjschen, Wikimedia Commons]

For a time-varying vector field, it is more proper to call the lines "pathlines" or "streaklines." ("Streamlines" are understood to generally apply to all visualizations of this variety.) Time-varying vector fields can be visualized by either method, using streamlines for snapshots of the field (i.e., holding time constant) or pathlines, when the field changes with time. Streamlines are a computer-based method that is very similar to experimental methods (such as helium-filled bubbles, for flow visualization. Figure 13 shows a dye-based visualization [39]). In Fig. 14, we see computer-generated streamlines.

There are more complex variants of the basic streamline. One version is called *stream-ribbons* [40]. Streamlines are one dimensional, as they do not have mathematical "thickness." They essentially track the trajectory of a particle through the flow field. Only a single vector can be tangent to a streamline at a particular location. By definition, a streamline is tangent to the flow field. It is not capable of showing twisting or swirling motion along its trajectory.

In some flow regimes, fluid may move along the direction of a streamline while circulating. A measure of this is called *streamwise vorticity*. Stream-ribbons are at

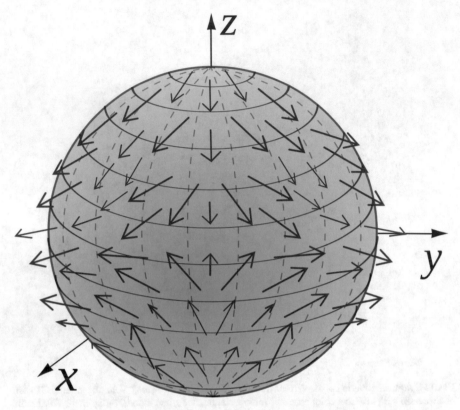

Fig. 12 Vector field visualized with a regular array of glyphs called a "Hedgehog" plot. From ["A Unit Sphere with Surface Vectors" by Cronholm144, Wikimedia Commons]

least two connected streamlines, drawn as a single object that resembles a ribbon. A twisting stream-ribbon can visually represent streamwise vorticity.

This method has application in weather visualization. Storms can exhibit helical regimes characterized by circulation about flow direction, for example, in convective updrafts.

Tensor Fields

As we have discussed, vectors are sets of individual numbers. The term *tensors*, when used synonymously as matrices, are just sets of vectors. "Tensor" may also be used to generalize the concept of any collection of numbers. So, with the latter usage of the term, scalars are zeroth-*order* tensors, vectors are first-*order* tensors, and matrices are second-*order* tensors. In this way, we can conceivably have nth-*order* tensors,

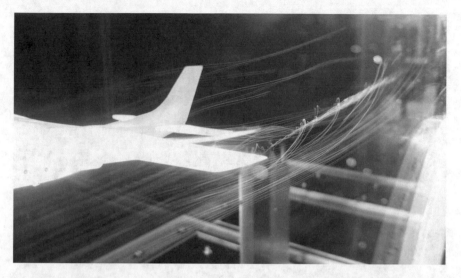

Fig. 13 Experimental flow visualization seen with helium bubbles. From ["A wind tunnel model of a Cessna 182 showing a wingtip vortex. Tested in the RPI (Rensselaer Polytechnic Institute) Subsonic Wind Tunnel" by Ben Frantz Dale, Wikimedia Commons]

i.e., collections of n-dimensional numerical constructs. A region of interest where there are known tensor quantities is a tensor field.

More recently, scientific visualization has begun to tackle matrix-level field data [41]. Data sets of this variety arise from a various disciplines. The most prominent tensor fields are in medical imaging and materials engineering. Again, we can attach glyphs to individual data points, or *advect* (integrate) streamlines using extractions of vectors from matrices.

Symmetric 3×3 matrices can be decomposed into three directions: that of maximum scaling and orthogonal directions to the principal eigenvector (i.e., maximum scaling direction) [42]. Symmetric matrices contain duplicate entries where row and column designations are reversed. Figure 15 shows glyphs of strain tensors in finite-element analysis visualization.

Figure 16 is a visualization of fiber tracks, in the human brain, as streamlines directed along orientation of water diffusion [43]. In this way, dendrites can be visualized using the eigenvectors of diffusion tensors, measured using *diffusion-tensor magnetic-resonance imaging* (DT-MRI).

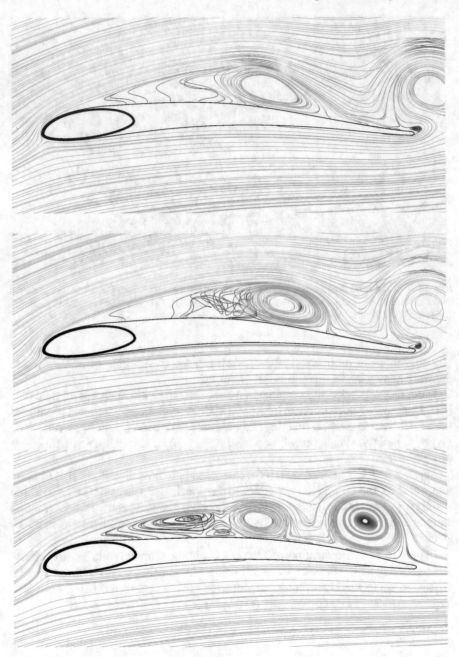

Fig. 14 Vortex-shedding from a flexible membrane airfoil. Reproduced with permission from [Elasto-Flexible Morphing Wing, Jonas Biehler, 2021]

Fig. 15 Visualization of strain tensor glyphs. Reproduced with permission from [Advanced visualization for Finite Elements Analysis in Virtual Reality Environments, Springer, 2008]

Summary

The creation of the electronic computer in the latter part of the twentieth century, and its integration with the Cathode Ray Tube (CRT) monitor, paved the way for our modern computer science-based practice of scientific visualization.

Finally, data collected from simulation and experiment could be stored and operated upon by automated computation, and numerical results displayed by algorithmic renditions of traditional and novel data-driven graphics. In this chapter, we introduced these fundamental algorithms and concepts that have enabled scientific visualization to be the field of computer science that it is today.

In the next chapter, we will discuss more recent progress, and in the last chapter we speculate on a few potential future enterprises for scientific visualization.

Fig. 16 Fiber tracks in the human brain. From [Thomas Schultz, Wikimedia Commons]

References

1. Robin, H.: The Scientific Image: From Cave to Computer. Abrams (1992)
2. Rovida, E.: Machines and Signs: A History of the Drawing of Machines, vol. 17. Springer Science & Business Media (2012)
3. Boyer, C.B.: History of Analytic Geometry. Courier Corporation (2012)
4. Turing, A.M.: On computable numbers, with an application to the Entscheidungsproblem. Proc. Lond. Math. Soc. **2**(1), 230–265 (1937)
5. Dyson, G.: Turing's Cathedral: The Origins of the Digital Universe. Pantheon (2012)
6. Von Neumann, J.: First draft of a report on the EDVAC. IEEE Ann. History Comput. **15**(4), 27–75 (1993)
7. Goldstine, H.H., Goldstine, A.: The Electronic Numerical Integrator and Computer (ENIAC). Math. Tables Aids Comput. **2**(15), 97–110 (1946)

8. Marcus, M., Akera, A.: Exploring the architecture of an early machine: the historical relevance of the ENIAC machine architecture. IEEE Ann History Comput. **18**(1), 17–24 (1996)
9. Swade, D., Babbage, C.: Difference Engine: Charles Babbage and The Quest to Build the First Computer. Viking Penguin (2001)
10. Williams, M.R.: The origins, uses, and fate of the EDVAC. IEEE Ann. History Comput. **15**(1), 22–38 (1993)
11. Neumann, J.V., Richtmyer, R.D.: A method for the numerical calculation of hydrodynamic shocks. J. Appl. Phys. **21**(3), 232–237 (1950)
12. Felker, J.H.: Performance of TRADIC transistor digital computer. In: Proceedings of the December 8–10, 1954, Eastern Joint Computer Conference: Design and Application of Small Digital Computers, pp. 46–49. ACM (1954)
13. Goldstine, H.H.: A History of Numerical Analysis from the 16th Through the 19th Century, vol. 2. Springer Science & Business Media (2012)
14. Chorin, A.J.: Numerical solution of the Navier–Stokes equations. Math. Comput. **22**(104), 745–762 (1968)
15. Taylor, C., Hood, P.: A numerical solution of the Navier–Stokes equations using the finite element technique. Comput. Fluids **1**(1), 73–100 (1973)
16. Peddie, J.: The History of Visual Magic in Computers. Springer, Berlin (2013)
17. Sutherland, I.E.: Sketchpad: a man-machine graphical communication system. Simulation **2**(5), R–3 (1964)
18. Sutherland, I.E.: Display Windowing by Clipping (1972). US Patent 3,639,736
19. Kajiya, J., Sutherland, I.: A Random-Access Video Frame Buffer, in Seminal Graphics: Pioneering Efforts That Shaped the Field. ACM Press (1975)
20. Blinn, J.F., Newell, M.E.: Texture and reflection in computer generated images. Commun. ACM **19**(10), 542–547 (1976)
21. Blinn, J.F.: Simulation of wrinkled surfaces. In: ACM SIGGRAPH Computer Graphics, vol. 12, pp. 286–292. ACM (1978)
22. Phong, B.T.: Illumination for computer generated pictures. Commun. ACM **18**(6), 311–317 (1975)
23. Lewis, M.: The New New Thing: A Silicon Valley Story. WW Norton & Company (1999)
24. Clark, J.H.: Hierarchical geometric models for visible surface algorithms. Commun. ACM **19**(10), 547–554 (1976)
25. Clark, J.: The geometry engine: a VLSI geometry system for graphics. In: Proceedings of SIGGRAPH'82. Citeseer (1982)
26. Luebke, D., Humphreys, G.: How GPUs work. Computer **40**(2), 96–100 (2007)
27. Dos Santos, S., Brodlie, K.: Gaining understanding of multivariate and multidimensional data through visualization. Comput. Graph. **28**(3), 311–325 (2004)
28. The HDF Group.: Hierarchical Data Format Version 5, 2000–2010. http://www.hdfgroup.org/HDF5
29. Alcock, L.: How to Think About Analysis. Oxford University Press, USA (2014)
30. Robinson, A.H.: The genealogy of the isopleth. Cartogr. J. **8**(1), 49–53 (1971)
31. Kumar, A., Verma, M.K.: Applicability of Taylor's hypothesis in thermally driven turbulence. R. Soc. Open Sci. **5**(4), 172152 (2018)
32. Kozlíková, B., Krone, M., Falk, M., Lindow, N., Baaden, M., Baum, D., Viola, I., Parulek, J., Hege, H.-C.: Visualization of biomolecular structures: state of the art revisited. In: Computer Graphics Forum, vol. 36, pp. 178–204. Wiley Online Library (2017)
33. Lorensen, W.E., Cline, H.E.: Marching cubes: a high resolution 3D surface construction algorithm. In: ACM Siggraph Computer Graphics, vol. 21, pp. 163–169. ACM (1987)
34. Drebin, R.A., Carpenter, L., Hanrahan, P.: Volume rendering. In: ACM Siggraph Computer Graphics, vol. 22, pp. 65–74. ACM (1988)
35. Pfister, H., Lorensen, B., Bajaj, C., Kindlmann, G., Schroeder, W., Avila, L.S., Raghu, K.M., Machiraju, R., Lee, J.: The transfer function bake-off. IEEE Comput. Graph. Appl. **21**(3), 16–22 (2001)

36. Laidlaw, D.H., Kirby, R.M., Jackson, C.D., Davidson, J.S., Miller, T.S., Da Silva, M., Warren, W.H., Tarr, M.J.: Comparing 2D vector field visualization methods: a user study. IEEE Trans. Vis. Comput. Graph. **11**(1), 59–70 (2005)
37. Telea, A.C.: Data Visualization: Principles and Practice. AK Peters/CRC Press (2014)
38. Runge, C.: Über die Numerische Auflösung von Differentialgleichungen. Mathematische Annalen **46**(2), 167–178 (1895)
39. Scarano, F., Ghaemi, S., Caridi, G.C.A., Bosbach, J., Dierksheide, U., Sciacchitano, A.: On the use of helium-filled soap bubbles for large-scale tomographic PIV in wind tunnel experiments. Exp. Fluids **56**(2), 42 (2015)
40. Hultquist, J.P.M.: Constructing stream surfaces in steady 3D vector fields. In: Proceedings of the 3rd Conference on Visualization'92, pp. 171–178. IEEE Computer Society Press (1992)
41. Kindlmann, G.: Superquadric tensor glyphs. In: Proceedings of the Sixth Joint Eurographics-IEEE TCVG conference on Visualization, pp. 147–154. Eurographics Association (2004)
42. Scherer, S., Wabner, M.: Advanced visualization for finite elements analysis in virtual reality environments. Int. J. Interact. Des. Manuf. (IJIDeM) **2**(3), 169–173 (2008)
43. Basser, P.J.: Fiber-tractography via diffusion tensor MRI (DT-MRI). In: Proceedings of the 6th Annual Meeting ISMRM, vol. 1226. Sydney, Australia (1998)

Recent Developments

Abstract Since the invention of the load-store computer, the direction of scientific visualization has been to increase coverage of scientific methodology and inquiry. Many advancements in modern scientific visualization have come from advances in computer hardware and rendering algorithms, applied to traditional topics such as chemical and biological macromolecular visualization. This chapter continues the journey into modern extensions of scientific visualization and uncertainty visualization, i.e., the use of large data sets for statistical visualization.

Biological and Chemical Processes

Molecular visualization—along with other forms of scientific visualization such as anatomical or medical imaging—was an early application. Its beginnings were discussed in prior chapters. However, continual advancements are being made. Looking at model visualization, we can see that current research papers are concerned with advancing rendering for large molecules. Part of this objective has been pursued using novel surface definitions. We focus on biomolecular visualization, as this has had the greatest progress.

Kozlikova et al. [1] have defined four overlapping types of visualization:

- static geometry,
- animation,
- intramolecular (within molecules), and
- intermolecular (between molecules).

These categories are depicted in Fig. 1. In this chapter, we limit our discussion to representational models, and methods for their rendering.

Traditional models were atomistic, either bond centric or van der Waals force based (union of all atomic radii). Bond-centric models are the familiar "ball-and-stick" variety. However, as traditional interactive rendering is triangle based, newer techniques have been proposed to render ellipsoidal primitives (in this case, as atoms) based on real-time ray casting. Gumhold et al. [2] originally presented a method for ellipsoid rendering for tensor fields, for fast interactive rendering. The method is

© The Author(s), under exclusive license to Springer Nature Switzerland AG 2022   79
B. E. Hollister and A. Pang, *A Concise Introduction to Scientific Visualization*,
https://doi.org/10.1007/978-3-030-86419-4_5

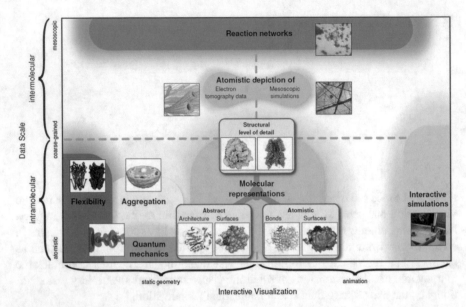

Fig. 1 Kozlikov's classification scheme of modern molecular visualization categories. From [Visualization of Biomolecular Structures: State of the Art Revisited, Wiley Online, 2016 Creative Commons Attribution License]

illustrated in Fig. 2. This has also enabled molecule rendering without tessellation of ellipsoidal primitives.

Further developments along this front have enabled faster rendering, and thus larger models consistent with macromolecular sizes. Chavent et al. created a GPU ray-casting algorithm employing hyperboloids as atomic bonds [3]. See Fig. 3.

Machine Learning in Scientific Visualization

One use of machine learning in scientific visualization has been as a filter in the visualization pipeline. Another has been to learn parameters for feature extraction by presentation of useful visualization examples [4].

Probability has worked its way as a fundamental aspect of data to be visualized directly [5]. We will talk more about probability and statistics for scientific visualization later, in the section on *uncertainty* visualization.

Machine learning is now used frequently in visualization. It can often provide *derived* features in scientific data; features low-level enough for human visual interpretation. Additionally, machine learning algorithms often require hyperparameters and values used to define the sensitivity of a machine learning algorithm.

Machine learning attempts to classify (or label) items of a data set. We consider two approaches: supervised and unsupervised. Where supervised learning algorithms

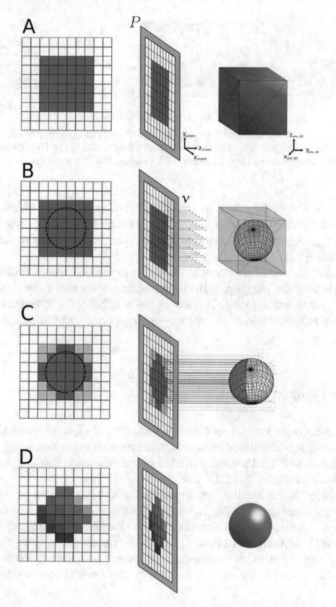

Fig. 2 Ellipsoidal splatting via GPU ray casting. Diagram from Chevent et al. **a** Bounding geometry made up of triangular primitives enclosing an ellipsoid. **b** Ray casting to determine pixels present in enclosing geometry. **c** "Splatting" of ellipsoid in screen-space (fragment shader). **d** Lighting applied in the fragment shader. Reproduced with permission from [GPU-Accelerated Atom and Dynamic Bond Visualization Using Hyperballs: A Unified Algorithm for Balls, Sticks, and Hyperboloids, Wiley Online, 2011 Free Access]

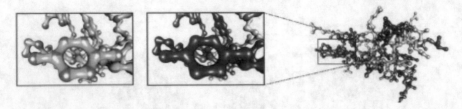

Fig. 3 Hyperboloid bonds rendered using GPU, from Chavent et al. Reproduced with permission from [GPU-Accelerated Atom and Dynamic Bond Visualization Using Hyperballs: A Unified Algorithm for Balls, Sticks, and Hyperboloids, Wiley Online, 2011 Free Access]

are trained on labeled data sets, unsupervised learning is not. In many instances, unsupervised learning is usually easier to understand in terms of how it reaches its decisions. Labeled data is often difficult or impossible to obtain for exploratory scientific visualization.

In this chapter, we constrain our discussion to various unsupervised learning algorithms in machine learning. In the final chapter, we see how deep learning (both supervised and re-enforcement) is opening the possibility for visualization of the mechanics of automated decision-making—with applications to scientific visualization.

Principle Component Analysis

Principal Component Analysis (PCA) is a standard technique in modern statistics. It is used for dimensionality reduction in data sets with many free variables (more than three). It can be used to view variation over a dimension that is a combination of orthogonal dimensions from the data.

PCA reduces dimensionality (number of variables) by finding an orthonormal set of axes for the same data set. Each new axis is a linear combination of the original variables and is sorted by how much each one can "explain" or account for the data. As such, the set of axes are optimized to find the maximum variances when data is projected onto them. PCA achieves dimensionality reduction by projecting high dimensionality data onto a few principal components (usually two or three).

Clustering

One form of unsupervised learning is classification by common features. Multiple algorithms exist for clustering along with different metrics for measuring similarity, e.g., Euclidean distance, density, etc. Clusters are another form of reduction of complexity in data. This is accomplished by collapsing records (entities in a data set) to a representative grouping. A few commonly used clustering algorithms are

- Density-based spatial clustering of applications with noisy data (DBSCAN) [6].
- k-means.
- Minimum spanning tree (MST).

Depending on the nature of the data entities, various combinations of parameters and algorithms might be used. Ultimately, visualization uses filtered data based on the categories of members (clusters) for rendering.

Uncertainty Visualization

The incorporation of uncertain data [7] is now an established visualization research and has been designated as a significant aspect of modern visual techniques. In contrast to crisp or certain data, uncertainty visualization strives to depict both data and its associated uncertainty. Visualization algorithms now provide provisions to account for data uncertainty. Related to this work is inclusion of statistical metrics for isosurface extraction and displaying explorations of parameters (for example, those used in simulation) for experimentation.

Probability Density Functions

An alternate way to consider statistical information is to include multiple values as a *probability density function* (PDF) and use methods for classifying them. This avenue of research began by extending techniques to work with PDFs [8]. Slice planes and clustering can be used to reduce the data dimension for rendering. Colormaps, glyphs, and deformations have been used to express summaries and related groups of data.

Ensemble Data

Many applications in physical sciences, engineering, statistics, risk assessment, decision science, etc. use *Monte Carlo* methods to model phenomena that are inherently uncertain. The input parameter space of the models is repeatedly sampled, and each sample set is solved using a deterministic model, to produce a possible outcome of the model. Each possible outcome is called a realization, and the collection of realizations from repeated runs is called an *ensemble*.

An everyday example is the weather forecast. Forecasts are usually obtained by running Monte Carlo simulations on a number of weather models. Each model may in turn be run with a set of input parameter whose values are drawn from a probability distribution associated with each parameter. An ensemble weather forecast may produce several fields such as temperature, humidity, pressure, and velocity. This gives rise to ensemble scalar fields and ensemble vector fields (EVF). Each data point is a distribution of values about the scalar or vector variable at each location and time. Hence, ensembles encode both the data and the uncertainty about the data.

It should be noted that spatially parallel scalar fields, containing values of separate and potentially uncorrelated measurements at the same location and time, are referred to in the aggregate as *multi-field* or *multivariate* fields. This is to distinguish them from vector fields, where the individual vector components constitute a single measurement of a quantity containing more than one dimension, such as in the case of velocity or force. Thus, we may have an ensemble of scalar fields, vector fields, tensor fields, or multivariate fields (containing any combination of scalar, vector, or tensor quantities).

Treatment of the different fields of an ensemble depends on the cardinality of the field. Ensemble scalar fields may be summarized using parametric statistics in certain situations. For example, the mean field can be used as a proxy for the ensemble scalar field, while the standard deviation field may be used as a representation for the uncertainty of the scalar field. This works when the distribution can be adequately characterized by parametric statistics [9]. However, this is not always the case. In situations where this assumption does not hold, non-parametric statistics may be computed and mapped visually.

Probabilistic Contours

Isosurface extraction shows characteristics of three-dimensional scalar data. Formulations of isocontours [10] allow for the display of positional uncertainty of isosurfaces colored by their level-crossing probability, as shown in Fig. 4. Rather than using isosurfaces to directly convey uncertainty in data, they can also be used to show shape and extent of clusters.

Statistical Medical Data

Fout and Ma [11] created an interactive tool to inspect uncertainty using a model that computes bounds on the uncertainty propagated by the volume rendering algorithm, shown in Fig. 5. Additional uncertainty visualization has been represented as transparency, glyphs, and color transfer.

Three-dimensional representations are very useful for geometric structure representation and providing context. However, the complexity of data usually requires multiple presentation types to enable better understanding [12]. Because of this, multi-window linked-view systems are popular for addressing uncertainty. The effect

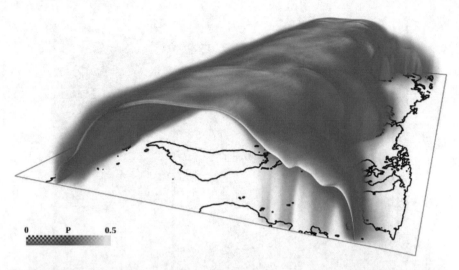

Fig. 4 Rendering from Pöthkow and Hege. Contour of 0^{circ} C in atmosphere with uncertainty. Colors indicate level-crossing probability, with warmer colors showing greater probability. Reproduced with permission from [Positional Uncertainty of Isocontours: Condition Analysis and Probabilistic Measures, IEEE, 2011]

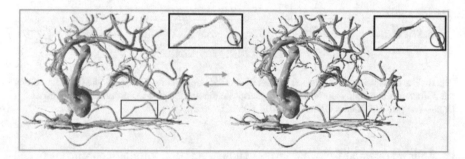

Fig. 5 Volume rendering of an aneurysm data set. Uncertain volume data produces multiple possible images. The two renderings shown here are the extremes of uncertainty, with the one on the left suggesting blood vessel restriction of the inset (Fout et al.). Reproduced with permission from [Fuzzy Volume Rendering, IEEE, 2012]

of changes to input parameters can be linked to alternative views such as parallel coordinates. This visualization contains multivariate data points which are represented as a piecewise linear curve through a set of axes for each variable arranged parallel to each other. An example of a parallel coordinates plot is shown in Fig. 6.

Maries et al. [13] proposed a visual comparison framework to explore correlation in neurological MRI volumes with patient mobility statistics, as a tool for early diagnosis and prevention. Their work focused on mobility issues related to neurological data from an elderly cohort using similar interactively linked views. A related work by Rosen et al. [14] proposed uncertainty visualization in a linked

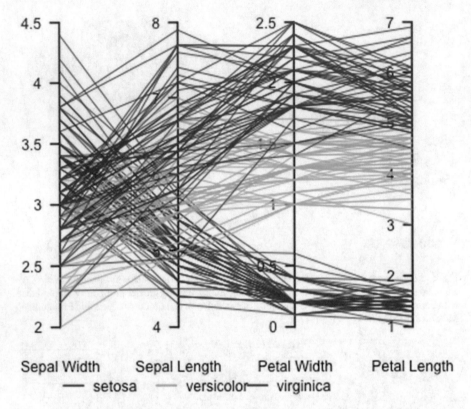

Fig. 6 The canonical "Iris" data set plotted in parallel coordinates. Each variable is labeled along the bottom of the graph. Reproduced with permission from [Parallel Coordinate Plot of Fisher Iris Data—JontyR, Wikimedia Commons, 2011 Public Domain]

view for myocardial ischemia inverse simulation data. Parallel coordinates linked with a three-dimensional heart model allow users to interactively view probabilistic ensemble data. See Fig. 7.

Uncertain Flow Data

Note that while some assumptions are imposed on the input parameters, e.g., Gaussian distribution, the EVF may not exhibit such properties. In fact, most of the interesting events happen when and where such assumptions fall apart. Otto et al. [15] examined uncertain vector field topology using Gaussian (normal bell-curve distribution) uncertainty.

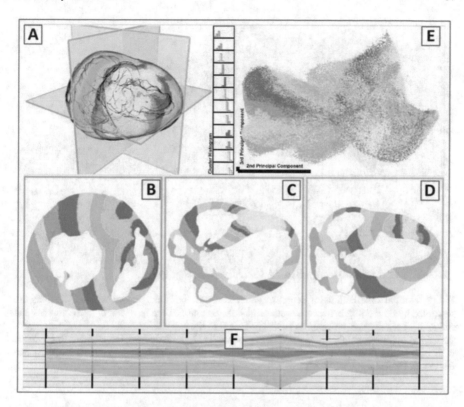

Fig. 7 The muView application. Caption and figure from Rosen et al.,"Three-dimensional view, B-D: Two-dimensional views, E: Feature space view of the principal component analysis of the PDFs, F: Parallel coordinates. Colored clustering of data points (k-means)." Reproduced with permission from [muView: A Visual Analysis System for Exploring Uncertainty in Myocardial Ischemia Simulations., Springer, 2016]

The affects of uncertain sources and sinks with normal distribution on regions flowing between them were first visualized. They also used a Monte Carlo streamline advection algorithm to produce flow paths. See Fig. 8.

Much work has been done to define and identify global features of flow fields for certain (or crisp) vector fields. Lagrangian Coherent Structures (LCS) are a broad class of feature identification for the fluid medium [16]. Perhaps the first notable example is the Finite-Time Lyapunov Exponent (FTLE) fields [17] for steady and unsteady vector field visualizations.

Generalization of LCS has been discussed in depth. Frameworks for flow field structure definition and visualization have been laid out by Salzbrunn et al. [18] There, the authors discuss pathline predicate definitions relevant for given investigations of flow phenomena.

Fig. 8 Normal distributions of sources, which are red, and sinks shown in blue. Regions are separated by flow going between a source and sink. Because of the uncertainty in the sources and sinks, these regions are uncertain as well. The height of the region shows the degree of certainty of the region, with full height being maximum probability for a distinct region of flow. Reproduced with permission from [Uncertain2D Vector Field Topology, The Eurographics Association and Blackwell Publishing Ltd., 2010]

A variance-based FTLE-like method for unsteady uncertain vector fields was first presented by Schneider et al. [19] This method, called Finite-Time Variance Analysis (FTVA), reports the maximum spatial second moment (variance) of particle destination, using the principal components of their covariance matrix as a result of initial uncertainty in the vector field. See Fig. 9.

To better understand the FTVA, let us look at the covariance matrix. This matrix is computed from the positions of the deposited particles at some finite time. For a two-dimensional vector field, we consider the x- and y-components of each particle's final position. Therefore, a covariance matrix comprised of the covariances for these positions would be two rows by two columns, or 2×2. The entries are the covariances of the random X and Y. (In order to compute the discrete covariances, it is assumed that each x, y pair occurs with equal probability.) The 2×2 covariance matrix is then

$$\begin{bmatrix} \sigma_{XX} & \sigma_{XY} \\ \sigma_{YX} & \sigma_{YY} \end{bmatrix}$$

where σ represents the covariance, and the subscripts X and Y are the random (x, y) particle positions. The diagonal of the covariance matrix simply contains the variances for the random variables.

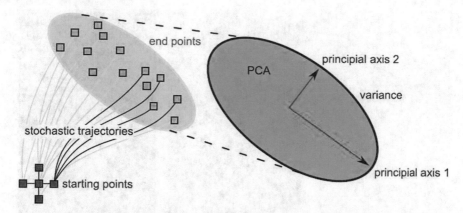

Fig. 9 Figure from Schneider et al. Stochastic integration from a starting point gives a distribution of end points due to uncertainty. A principal component analysis of the start and end point distributions provides information about the maximum amount of "stretching," an analogy to physical fluid expansion but instead over an ensemble of possibilities with increasing variation. Reproduced with permission from [A Variance Based FTLE-Like Method For Unsteady Uncertain Vector Fields, Springer, 2012]

By determining the maximum eigenvalue and its corresponding eigenvector, we find the principal axis and degree of variance along with it, which is also the largest uni-directional variance in particle position. Note that an eigenvalue is a scalar that acts the same as a matrix when multiplied by an eigenvalue's matching eigenvector. Numerical methods are often used for the calculation of eigenvector-eigenvalue pairs.

Hummel et al. [20] were the first, however, to apply FTVA to address EVF visualization. They use both individual and joint variance colormappings to discern sources of uncertainty and reliability in trend analysis. See Fig. 10. Their paper also used a Minimum Spanning Tree (MST) to detect and visualize trends (clusters) in particle destinations.

Guo et al. [21] outlined a framework to provide an interactive assessment of ensemble variation. They call their system eFLAA (ensemble Flow Line Advection and Analysis). See Fig. 11. They present a novel parallel computation for calculating streamline spatial difference over an ensemble and then visualizing the differences. They compute various features of their ensembles (e.g., carbon dioxide concentration) along streamlines whose variation meets a given threshold.

Boxplots have been extended from points to curves such as streamlines [22]. Mirzargar et al. apply their method to visualize ensemble streamlines and hurricane track data. See Fig. 12.

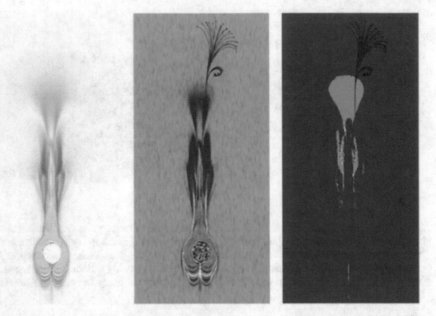

Fig. 10 Variances in ensemble flow around a cylinder in two dimensions. Left-most image shows naive joint (ensemble) variance from the FTVA. Center and right-most images depict individual variation as well with joint variance, with white areas showing high values for both. Trends are most reliably the source of joint variance when individual variance is also low according to Hummel et al. Reproduced with permission from [Comparative Visual Analysis of Lagrangian Transport in CFD Ensembles, IEEE, 2013]

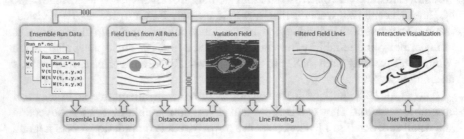

Fig. 11 Guo's distance computation and visualization of ensemble flow data. Reproduced with permission from [Coupled Ensemble Flow Line Advection and Analysis, IEEE, 2013]

As a last example of uncertain flow visualization, we can see in Fig. 13 an extension which allows for trend visualization in flow paths called *Streamline Variability Plots* [23].

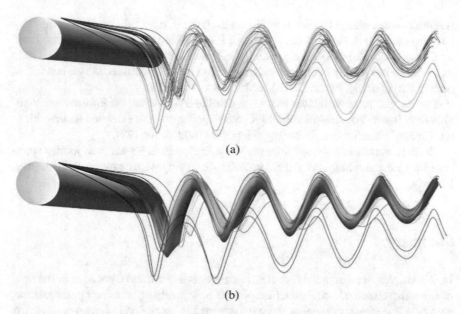

(a)

(b)

Fig. 12 The *Curve Boxplot*. **a** Shows a regular ensemble of streamlines, and **b** shows a curve boxplot with outlier flow paths. Reproduced with permission from [Curve Boxplot: Generalization of Boxplot for Ensembles of Curves, IEEE, 2014]

Fig. 13 *Streamline Variability Plots*, which visualize clusters of closely related flow paths in an ensemble. Reproduced with permission from [Streamline Variability Plots, IEEE, 2016]

Conferences

As scientific visualization has developed, dedicated conferences are now responsible for most of the research presentations on current advancement. There are presently three top-tier conferences: IEEE Visualization Conference,[1] IEEE Pacific Visualization (Pacific Vis),[2] and Eurovis.[3]

Interestingly, IEEE VIS grew out of the American Computing Society's (ACM) Special Interest Group for Graphics (or SIGGRAPH) [24], where seminal work in

---

[1] http://ieeevis.org

[2] https://www.facebook.com/PacificVis/

[3] https://www.eurovis.org

scientific visualization has been published in the past, e.g., "Volume Rendering" by Drebin et al. [25] in 1988, "Imaging Vector Fields Using Line Integral Convolution" by Cabral et al. [26] in 1993, etc. The IEEE Visualization Conference was first held in 1990 and has mostly been held in cities across the U.S. until recently when it was held in Paris (2014), Berlin (2018), and Vancouver (2019).

Pacific Vis' prior name was the Asian-Pacific Symposium on Information Visualization. It was first held in 2008 in Kyoto City, Japan, and extended its breadth to all forms of visualization, including scientific visualization [27].

Eurovis started as a Symposium on Visualization in 1999 and was jointly sponsored by Eurographics and IEEE. In 2005, the symposium saw the transition to Eurovis.

Summary

In this chapter, we discussed recent advancements and current directions in the field of scientific visualization. As the discipline is now firmly a part of computer science, particularly computer graphics, its development has hinged on improvements across the field of computing. The interested reader is encouraged to pursue many of the referenced papers and surveys, for more examples.

References

1. Kozlíková, B., Krone, M., Falk, M., Lindow, N., Baaden, M., Baum, D., Viola, I., Parulek, J., Hege, H.-C.: Visualization of biomolecular structures: state of the art revisited. In: Computer Graphics Forum, vol 36, pp. 178–204. Wiley Online Library (2017)
2. Gumhold, S.: Splatting illuminated ellipsoids with depth correction. In: VMV, pp. 245–252 (2003)
3. Chavent, M., Vanel, A., Tek, A., Levy, B., Robert, S., Raffin, B., Baaden, M.: GPU-accelerated atom and dynamic bond visualization using hyperballs: a unified algorithm for balls, sticks, and hyperboloids. J. Comput. Chem. 32(13), 2924–2935 (2011)
4. Ma, K.-L.: Machine learning to boost the next generation of visualization technology. IEEE Comput. Graph. Appl. 27(5), 6–9 (2007)
5. Johnson, C.R.: Top scientific visualization research problems. Computer Graphics and Applications. IEEE (2004)
6. Ester, M., Kriegel, H.-P., Sander, J., Xiaowei, X.: A density-based algorithm for discovering clusters in large spatial databases with noise. Kdd 96, 226–231 (1996)
7. Johnson, C.R., Sanderson, A.R.: A next step: visualizing errors and uncertainty. Comput. Graph. Appl. IEEE 23(5), 6–10 (2003)
8. Luo, A., Kao, D., Pang, A.: Visualizing spatial distribution data sets. In: VisSym (2003)
9. Love, A.L., Pang, A., Kao, D.L.: Visualizing spatial multivalue data. IEEE Comput. Graph. Appl. 25(3), 69–79 (2005)
10. Pöthkow, K., Hege, H.-C.: Positional uncertainty of isocontours: condition analysis and probabilistic measures. Vis. Comput. Graph. IEEE
11. Fout, N., Ma, K.-L.: Visualization and Computer Graphics, IEEE Transactions on Fuzzy, vol. Rendering (2012)

12. Berger, W., Piringer, H., Filzmoser, P., Gröller. E.: Uncertainty-aware exploration of continuous parameter spaces using multivariate prediction. In: Computer Graphics Forum
13. Maries, A., Mays, M., Hunt, M., Wong, K.F., Layton, W., Boudreau, R., Rosano, C., Marai, G.E.: Grace: a visual comparison framework for integrated spatial and non-spatial geriatric data. IEEE Trans. Vis. Comput. Graph. (2013)
14. Rosen, P., Burton, B., Potter, K., Johnson, C.R.: muView: A Visual Analysis System for Exploring Uncertainty in Myocardial Ischemia Simulations (2016)
15. Otto, M., Germer, T., Hege, H.-C., Theisel, H.: Uncertain 2D vector field topology. In: Computer Graphics Forum (2010)
16. Peacock, T., Haller, G.: Lagrangian coherent structures: the hidden skeleton of fluid flows. Phys. Today (2013)
17. Haller, G.: Distinguished material surfaces and coherent structures in three-dimensional fluid flows. Phys. D: Nonlinear Phenom. (2001)
18. Salzbrunn, T., Garth, C., Scheuermann, G., Meyer, J.: Pathline predicates and unsteady flow structures. Vis. Comput. (2008)
19. Schneider, D., Fuhrmann, J., Reich, W., Scheuermann, G.: A variance based ftle-like method for unsteady uncertain vector fields. In: Topological Methods in Data Analysis and Visualization II, pp. 255–268. Springer, Berlin (2012)
20. Hummel, M., Obermaier, H., Garth, C., Joy, K.: Comparative visual analysis of lagrangian transport in CFD ensembles. Vis. Comput. Graph. (2013)
21. Guo, H., Yuan, X., Huang, J., Zhu, X.: Coupled ensemble flow line advection and analysis. Vis. Comput. Graph. (2013)
22. Mirzargar, M., Whitaker, R., Kirby, R.: Generalization of boxplot for ensembles of curves. Curve Boxplot (2014)
23. Ferstl, F., Bürger, K., Westermann, R.: Streamline variability plots for characterizing the uncertainty in vector field ensembles. IEEE Trans. Vis. Comput. Graph. $22(1)$, 767–776 (2015)
24. Salzman, D., Von Neumann, J.: Visualization in scientific computing: summary of an NSF-sponsored panel report on graphics, image processing, and workstations. Int. J. Supercomput. Appl. $1(4)$, 106–108 (1987)
25. Drebin, R.A., Carpenter, L., Hanrahan. P.: Volume rendering. In: ACM Siggraph Computer Graphics, vol. 22, pp. 65–74. ACM (1988)
26. Cabral B., Leedom, L.C.: Imaging vector fields using line integral convolution. In: Proceedings of the 20th Annual Conference on Computer Graphics and Interactive Techniques, pp. 263–270 (1993)
27. Kyoto University.: IEEE VGTC Pacific Visualization Symposium (2008). http://www.viz. media.kyoto-u.ac.jp/conf/pvis2008/index.html

The Future

Abstract A significant question raised is, "Does scientific visualization have continued utility as originally practiced?" We will explore the integration of scientific visualization into aspects of education and professional collaboration. Additionally, researchers are becoming more reliant on artificial intelligence to make sense of their data. In contrast to scientific visualization, artificial intelligence is often opaque to the human observer. There are efforts toward explainable deep learning, as the algorithms are not transparent in their predictions. Scientific visualization has similar goals so that one can "drill down" and get more information about the data and perhaps why the visualization is the way that it is presented.

In this chapter, predictions lie within the next 5 years. There will be more cross-over (and cross-fertilization) between scientific visualization and machine learning (e.g., as many problems in scientific visualization deal with engineering and science in particular). Physics-aware machine learning models may be irrelevant; uncertainty in machine learning models especially when there is noise in the training data and how that impacts a model's accuracy.

Fidelity, Speed, and Content

One way of extrapolating progress is simply to "up-size" the present. Thus, higher resolution, smoother immersive visuals, larger, and instantaneously delivered streaming data are a few of the published prognostications [1]. Unfortunately, this approach often fails when looking back on earlier predictions [2].

Critical advances, though, do not happen linearly or smoothly but are discontinuous by their nature. Their causes are many, because of physical (or practical) impediments to technological advancement. Often, significant leaps result from the amalgamation and realignment of standard approaches. Both paths, propelled by motivation or necessity, can guide us in approximating potential advancements. Such avenues are discussed in this section.

B. E. Hollister and A. Pang, *A Concise Introduction to Scientific Visualization*,
https://doi.org/10.1007/978-3-030-86419-4_6

Big Data

"Big Data" is marked by the definition of data having the following attributions: volume, variety, velocity, and veracity [3]. While large data sets have come to the public awareness through companies such as Google and Facebook, scientific computing and visualization has long dealt with large amounts of data and the problems it presents for visualization.

A new direction [4] in scientific "big data" is its integration (visualization) into print articles with cross-reference between data sources, e.g., molecular ensembles [5]. See Fig. 1 for an example. These approaches are expected to become more prevalent, over a larger variety of scientific disciplines.

Distributed Collaboration

Rhyne defines scientific visualization as "computationally intense visual thinking" [6]. She also predicts "global virtual environments for interactively exploring and immersing ourselves in 3D representations of scientific phenomena." This is becoming more possible with devices such as MicroSoft's HoloLens. See Fig. 2.

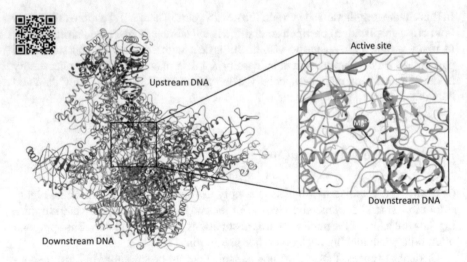

Fig. 1 Inclusion of large interactive biomolecules in printed articles via QR Code. Users scan code and automatically download an interactive figure to a local visualization program. Reproduced with permission from [Augmented Reality in Scientific Publications—Taking the Visualization of 3D Structures to the Next Level, American Chemical Society, 2018]

Fig. 2 Collaborative visualization via augmented reality. Reproduced with permission from [NEXT16 Hololens Demo 2—NEXT Conference from Berlin, Wikimedia Commons, 2016 Creative Commons Attribution 2.0 Generic]

Another aspect here is the use of open source for scientific visualization. Rhyne also states, "One of the concepts that we will continually rediscover is that the compelling message and meaning associated with a scientific visualization is open source." The Visualization Toolkit, or VTK, which is an open-source project has been instrumental in bringing newer methods into common use [7].

Scientific Visualization for Education

Limniou et al. [8] discuss using a Cave Automatic Virtual Environment (CAVE) as a way to describe three-dimensional chemical structures and their paper reports improved student understanding of molecular mechanics. Simulation of actual laboratory practice in a virtual environment has been studied by Martinez-Jimenez et al. [9] Their software allows teachers to provide tests as part of the simulations. Both of these publications address presentation and training. X-ray crystallography, a technique requiring expensive and potentially dangerous equipment, is just beginning to use virtual reality in an educational setting (Fig. 3) [10, 11].

Zheng et al. [12] suggest there is an epidemic of unfamiliarity with the underlying nature of biomolecular structural information. Tellingly, Zheng states, "X-ray crystallography will find its future application in drug discovery by the development

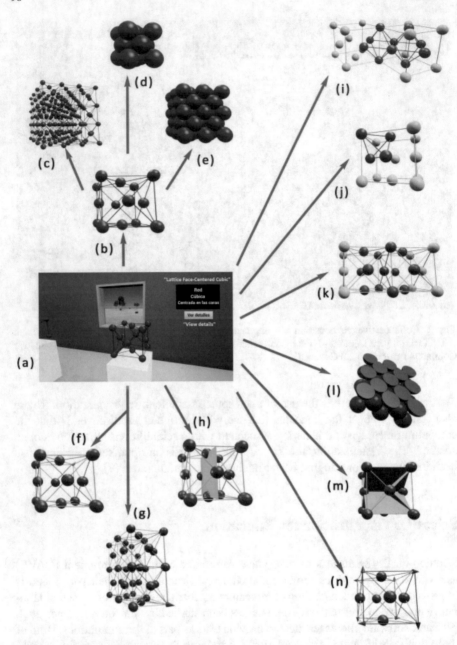

Fig. 3 Virtual reality training in chemistry. "Different views of a material structure: **a** lattice FCC; **b** expanded unit cell; **c** expanded set; **d** unit cell in CPK mode; **e** super set in CPK mode; **f** crystallographic direction [011]; **g** crystallographic direction [210]; **h** crystallographic plane [020]; **i** octahedral interstitial sites (voids); **j** tetrahedral interstitial site (void); **k** coordination index; **l** cross section; **m** family of planes 110; and **n** family of directions <100>." Reproduced with permission from [Virtual and Augmented Reality Environments to Learn the Fundamentals of Crystallography, Crystals, 2020 Open Access]

of specific tools that would allow realistic interpretation of the outcome coordinates and/or support testing of these hypotheses." These concerns should be addressed early in the education of future chemists, but because of the inherent difficulty in teaching many types of physical methods at the undergraduate level, there is a real need for a substitute involving novel STEM educational methods. Not all institutions have either the expertise or requisite funding.

O'Hara et al. [13] present findings involving undergraduate computer science students in open-source software projects. Software engineering best practices and involvement with large code bases are the useful by-products of open source in education. Shindelin et al. [14] demonstrate Fiji, an open-source image processing application for biological imaging. They discuss how their software provides a fertile area for collaboration between students of computer science and biology. However, despite these studies and the importance of involving undergraduate computer science students in cross-disciplinary research and open-source software development (essential in the area of scientific visualization) [15], no current applications address comprehensive crystallography visualization for STEM education.

Scientific Visualization and Artificial Intelligence

Perhaps the most disruptive aspect for the future practice of scientific visualization will be increasing reliance on artificial intelligence (AI). Machine learning is already used for various stages in the visualization pipeline, as we saw from the previous chapter. But a question arises, "Will more ambitious forms of AI, such as deep learning, be used effectively for scientific visualization—or are the two incompatible?"

Deep Learning

A key problem with deep learning is interpreting the mappings created by trained artificial neural networks (ANN). While classes of data may be effectively predicted with such networks, the hidden-layer neurons provide no theoretical model (analytical function). See Fig. 4. If ANNs are used in the visualization pipeline, it is only to understand their final output relative to other input data points.

Visualization using dimensional reduction, such as t-Distributed Stochastic Neighbor Embedding (t-SNE) [16], may provide insight into derived features. See Fig. 5 for an example of t-SNE. More broadly, high-dimensional data can be visualized with such dimensional filtering, including pathlines, molecular coordinates, or general network connections.

Fig. 4 Simplified fully connected ANN with one hidden layer and nine neurons. From [Wiso, Wikimedia Commons]

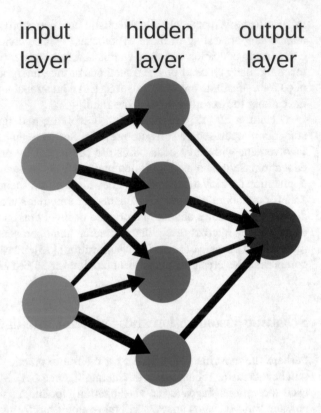

input layer   hidden layer   output layer

Bi-Directional Scattering Distribution Functions Using ANNs

Traditionally, BSDFs (Fig. 6) are first measured by automated gonioreflectometers, and then fit to either analytical functions or estimated by consistent physical models [17–19].

A speculative application of deep learning in conjunction with scientific visualization is formulating analytical models of Bi-Directional Reflectance or Scattering Distribution Functions (BRDF and BSDF, respectively) from radiometric data. Vidaurre et al. [20] match BRDF models from data but do not synthesize either BRDF or BSDF.

With better understanding of hidden-layer "intrinsic" features via visualization, BSDF may be theoretically derived from ANNs trained on scattering data.

Fig. 5 High-dimensional clusters visualized with t-SNE projected to two dimensions. Handwritten numbers (MNIST) data set (from Van der Maaten, et al.). Reproduced with permission from [Visualizing Data Using t-SNE, Journal of Machine Learning Research, 2008 Open Access]

Fig. 6 Measured BRDF from finished wood. Reflection intensities shown for the hemisphere over numbered points on surface (left). Such data is fit by human researchers, but this may change using ANNs and multidimensional visualization of hidden layers. Reproduced with permission from [Measuring and Modeling the Appearance of Finished Wood, ACM, 2005]

Summary

We have touched on a few possible paths scientific visualization may pursue in the future. Research will be based on recombination of current methods together with applications yet envisaged. In some instances, scientific visualization will serve as a replacement for real-world laboratory equipment in education. In others, multidimensional visualization of deep learning models may contribute to fundamental computer graphics knowledge.

Most prognostications in this chapter have been within the 5-year time horizon. However, beyond 5 years we see scientific visualization being used by an expanding user base (not only a fairly elite/small crowd of scientists and engineers). Expanding scientific visualization's user base will open up a host of new research problems. For example, commercial space travel might become a reality within our lifetimes, and there may be applications within that realm.

References

1. Olshannikova, E., Ometov, A., Koucheryavy, Y., Olsson, T.: Visualizing big data with augmented and virtual reality: challenges and research agenda. J. Big Data **2**(1), 1–27 (2015)
2. Brodlie, K.: Scientific visualization-past, present and future. Nucl. Instrum. Methods Phys. Res. Sect. A: Accel. Spectrometers Detect. Assoc. Equip. **354**(1), 104–111 (1995)
3. Holmes, D.E.: Big Data: A Very Short Introduction. Oxford University Press (2017)
4. Wolle, P., Mueller, M.P., Rauh, D.: Augmented Reality in Scientific Publications—Taking the Visualization of 3D Structures to the Next Level (2018)
5. Heinrich, J., Krone, M., O'Donoghue, S.I., Weiskopf, D.: Visualising intrinsic disorder and conformational variation in protein ensembles. Faraday Discuss. **169**, 179–193 (2014)
6. Rhyne, T.-M.: Scientific visualization in the next millennium. IEEE Comput. Graph. Appl. **20**(1), 20–21 (2000)
7. Hanwell, M.D., Martin, K.M., Chaudhary, A., Avila, L.S.: The Visualization Toolkit (VTK): rewriting the rendering code for modern graphics cards. SoftwareX **1**, 9–12 (2015)
8. Limniou, M., Roberts, D., Papadopoulos, N.: Full immersive virtual environment CAVE TM in chemistry education. Comput. Educ. **51**(2), 584–593 (2008)
9. Martínez-Jiménez, P., Pontes-Pedrajas, A., Polo, J., Climent-Bellido, M.S.: Learning in chemistry with virtual laboratories. J. Chem. Educ. **80**(3), 346–352 (2003)
10. Ratamero, E.M., Bellini, D., Dowson, C.G., Römer, R.A.: Touching proteins with virtual bare hands. J. Comput.-Aided Mol. Des. **32**(6), 703–709 (2018)
11. Extremera, J., Vergara, D., Dávila, L.P., Rubio, M.P.: Virtual and augmented reality environments to learn the fundamentals of crystallography. Crystals **10**(6), 456 (2020)
12. Zheng, H., Hou, J., Zimmerman, M.D., Wlodawer, A., Minor, W.: The future of crystallography in drug discovery. Expert Opin. Drug Discov. **9**(2), 125–137 (2014)
13. O'Hara, K.J., Kay, J.S.: Open source software and computer science education. J. Comput. Sci. Colleges **18**(3), 1–7 (2003)
14. Schindelin, J., Arganda-Carreras, I., Frise, E., Kaynig, V., Longair, M., Pietzsch, T., Preibisch, S., Rueden, C., Saalfeld, S., Schmid, B. et al.: Fiji: An open-source platform for biological-image analysis. Nat. Methods **9**(7), 676–682 (2012)
15. Shneiderman, B.: The New ABCs of Research: Achieving Breakthrough Collaborations. Oxford University Press (2016)
16. Van der Maaten, L., Hinton, G.: Visualizing data using t-SNE. J. Mach. Learn. Res. **9**(11) (2008)
17. Matusik, W., Pfister, H., Brand, M., McMillan, L.: Efficient Isotropic BRDF Measurement (2003)
18. Ward, G.J.: Measuring and modeling anisotropic reflection. In: Proceedings of the 19th Annual Conference on Computer Graphics and Interactive Techniques, pp. 265–272 (1992)
19. Marschner, S.R., Westin, S.H., Arbree, A., Moon, J.T.: Measuring and modeling the appearance of finished wood. In: ACM SIGGRAPH 2005 Papers, pp. 727–734 (2005)
20. Vidaurre, R., Casas, D., Garces, E., Lopez-Moreno, J.: BRDF estimation of complex materials with nested learning. In: 2019 IEEE Winter Conference on Applications of Computer Vision (WACV), pp. 1347–1356. IEEE (2019)

Index

© The Editor(s) (if applicable) and The Author(s), under exclusive license to Springer
Nature Switzerland AG 2022
B. E. Hollister and A. Pang, *A Concise Introduction to Scientific Visualization*,
https://doi.org/10.1007/978-3-030-86419-4

Printed in the United States
by Baker & Taylor Publisher Services